Management and Development of Communications Networks

D Crawford
Associate Consultant
The National Computing Centre Limited

NCC Blackwell
MANCHESTER • OXFORD

British Library Cataloguing in Publication Data

Crawford, D.
Management and development of communications networks
I. Title
001.64404068

ISBN 1-85554-148-3

First published in 1992 by:

NCC Blackwell Ltd, 108 Cowley Road, Oxford OX4 1JF, England.

Editorial Office: The National Computing Centre Limited, Oxford House, Oxford Road, Manchester M1 7ED, England.

Typeset in 11pt Bookman by Typesetting Department, The National Computing Centre Limited, Oxford Road, Manchester M1 7ED and printed by Bookcraft (Bath) Ltd.

ISBN 1-85554-148-3

Acknowledgements

NCC would like to thank the following organisations for their help and assistance in the development of this report:

British Telecom

Mercury Communications

Telematics International

Olivetti

Plessey-Telenet

Case Communications

Infotron Systems

ICL

Timeplex

Racal-Milgo

BBN

Strathclyde Police

Greenalls

The Communications Insight Series

This is one of a series of reports from a two year Communications Programme which has been funded by the Department of Trade and Industry and British Telecom, and undertaken by NCC's Communication and Security Division. The objectives of the Programme, which runs until May 1992, are to:

- raise levels of awareness within small to medium enterprises of the new and developing communications technologies now available;
- promote the effective uptake of these technologies.

The Communications Insight Series has been produced to provide senior executives, IT managers, telecommunications managers and technical support staff with:

- a state-of-the-art perspective of the new communications technologies;
- details of innovative ways in which the new technologies may be applied;
- detailed investigations of how they can be successfully integrated with existing technologies.

The reports which make up the series are:

- *Managers Guide to ISDN Planning and Networking Products*
- *Management and Development of Communications Networks*
- *Mobile Data Communications - Choices and Opportunities*
- *ISDN Applications*
- *ISDN Terminal Products Profile*

Contents

Author's biography

David Crawford is an NCC Associate Consultant with over twenty years experience in the IT industry. As an independent consultant with considerable experience in developing large communications networks (particularly in financial systems), his expertise has been applied to a wide range of communications technologies to solve business problems. In the course of developing various networks, he has dealt with the many facets of systems integration as well as with the problems of managing large networks containing a wide variety of equipment and media.

In recent years he has worked within the development field on X.25 based networks and on systems employing OSI standards. More recently, as an NCC Associate Consultant, he has conducted strategy and technical studies for NCC and its clients.

One of his major interests is the promotion to users of employing a structured approach to providing a network management service that is also quality assured.

Introduction

Background

This is one of a series of management reports developed as part of an NCC Communications programme, which is sponsored by the Department of Trade and Industry and British Telecom.

The purpose is to promote greater awareness of what telecommunications technology can provide. The role of telecommunications networks is of fundamental importance in achieving business objectives in a large number of organisations. A high level of performance is expected and it is critical that networks are managed successfully. The introduction of packet switching and the X.25 interface into the communications network, as well as Local Area Networks (LANs), can add to the complexity of management. Solving any additional management difficulties is vital if the connectivity, routeing and switching functions are to provide effective solutions to the problems posed by the business requirements. Naturally, this will only be achieved if the requirements are clearly specified and analysed, and the service developed to satisfy them.

Many existing networks have become complex, so it is essential that quantitative measurements are made to determine that the level of service being provided is what is required by the users. The Network Manager must also determine what this level of service costs, and thus cost the business applications it supports. The following questions illustrate the type of issues that network management must address:

- What is the purpose of the network?
- Are its boundaries clearly defined?
- Are the Network Services clearly defined?
- Are there Service Level Agreements with users?
- Are there Level of Service Contracts with suppliers?
- Are Network Service performance and Equipment performance measured?
- How much does the network cost to run?
- What standards are used to measure quality of service?

The answers suggested in this report are based on the writer's experiences of developing and running communications networks.

Very few organisations are able to build a network from scratch. Nearly all will have an existing system of some kind, as well as an organisational structure and methods for managing it. Network developments should fit into the structure required to manage them, and, even if a development is massive, the organisational structure should be able to handle it, perhaps with additional staff and extra skills.

In this report, network management responsibilities are divided into four groups:

- operational control;
- administration;

- analysis;
- capacity planning;

so that related tasks are collected together. This may be also a convenient way of structuring staff functions, although the number of staff required for different sizes of network may be a more influential factor. This system of responsibility grouping has been established in practice, as well as being used in other reports and books on network management.

For network management to achieve the quality of service objectives in the four groups listed above requires:

- procedures to cover all network management activities and define responsibilities;

- support systems, for example, for fault management, performance analysis, inventory management, network utilisation and response time monitoring;

- human resources with specific skills.

This report provides technical and functional information on a number of X.25 products, and provides only a sample of the range of facilities available. There is information on switches, PADs, Gateways and Protocol Converters, as well as details of the facilities offered by the major service suppliers.

Unfortunately, network costs are not limited to the initial purchase the X.25 products and services, so there is also information on the other items that need to be included when calculating the cost of providing service to the user. The costs of lines and modems are fairly obvious, whilst the cost of network management is less tangible.

Although this report is structured around X.25 systems, the network management section could be applied to any network. The intention is to start with the assessment of business requirements and then to proceed to the evaluation of systems, from testing and implementation through to live running, assessing costs at all stages. The purpose is to provide advice and recommendations on any relevant aspects of development and running networks, especially those that should help to improve the chances of success in network management.

The report is not intended to be highly technical, although some technical information has been introduced in relation to specific network management techniques.

Summary of objectives

- To provide information on examples of X.25 product functions and services currently available.
- To advise on the implementation and running of networks.
- To recommend the support systems required in order that network management can meet tangible quality of service objectives.
- To advise on network management structure and to define the activities that should take place in the different parts of the structure.

Audience

- Those responsible for managing the development and operation of networks.

- Those with an interest in what X.25 products and services can be provided, for business applications.

- Those interested assessing the real costs of providing a network service.

- Those interested in organisational structures for network management and the activities that have to be carried out to provide the service.

1

X.25, other standards and their use

1.1 Summary

This chapter covers the following:

- the role that packet switched networks can play in an organisation, their importance both as a stable interfacing standard and because of the available functions, for example switching, multiplexing;

- an overview of the structure of X.25 and its role in packet switching;

- the relationship between X.25 and other OSI (Open Systems Interconnection) Lower and Transport layer standards;

- the importance of open system standards to organisations when procuring equipment, for interworking between systems and for encouraging developments, by suppliers, in communications technology.

1.2 Packet switching choice

X.25 and packet switching products have reached maturity, with

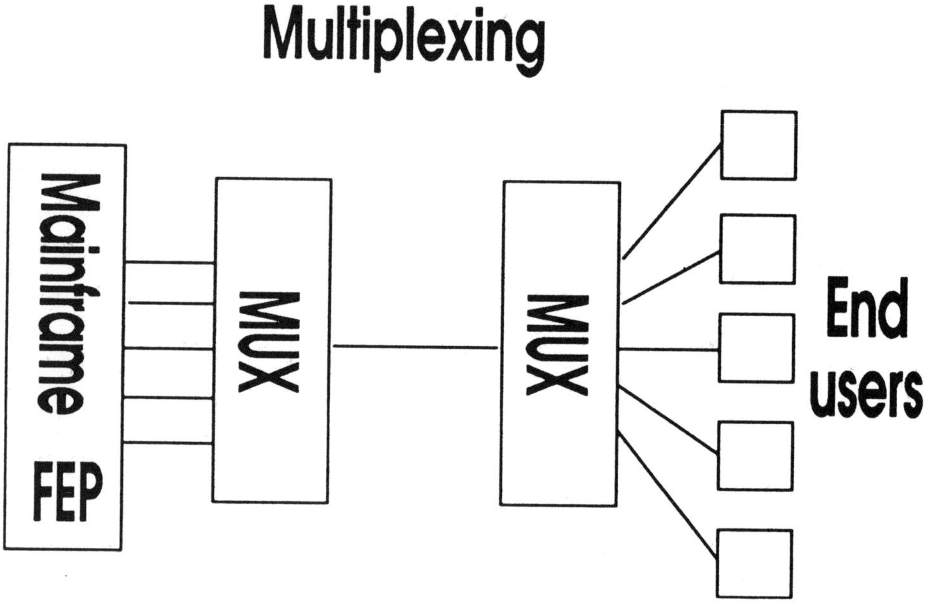

Figure 1 Multiplexing

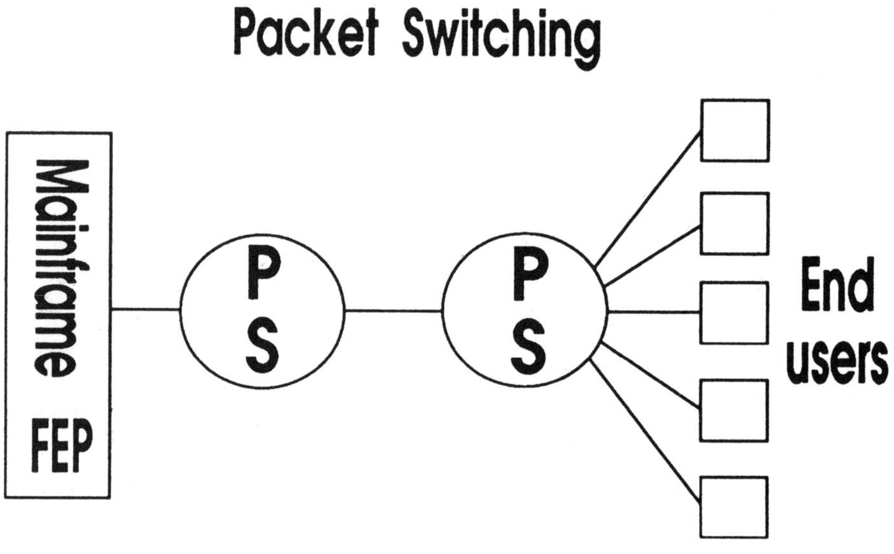

Figure 2 Packet Switching

many products now available for wide area networking. X.25 support on end-user systems is now commonplace. LAN Routers, with an X.25 gateway, can be connected to packet switched network to give LAN users a wide range of potential access points, for example to other LANs or to mainframes. X.25 is a key standard within OSI (Open Systems Interconnection), but it should not be assumed that OSI and X.25 is the only choice for providing a flexible and open network; the use of the Internet protocol, TCP/IP, is also widespread.

The inherent multiplexing capability of X.25 can provide concentration of data from a number of end-points on to a single physical link, e.g. when connecting terminals to a central mainframe. The multiplexing capability enables designers to estimate numbers of physical connections to a mainframe/FEP based on data volumes, rather than on physical user end-points as with time division and statistical multiplexing (*see* Figure 1 and Figure 2). The overwhelming majority of systems would only require a small number of mainframe connections at 64 kbits/sec. This reduces the FEP cost, as fewer ports are required.

Other cost differences arise from:

- the requirements being met by a more expensive type of network;

- the physical management requirements of the network, particularly for fault detection. The cost will depend on the functions available as standard in the purchased system. This may be dependent on product rather than on network type (*see* Figure 3).

Switching in a hierarchical type of system may, on the surface, seem to be of little importance. However, to minimise the risk of system loss, the configuration is likely to include a back-up mainframe and redundant network components. The inherent switching and multiplexing of packet switching make these requirements relatively simple to configure. As discussed above, the number of connections to the back-up mainframe should be low.

Managing the physical network (fault finding)

Fault finding on media, hardware & software in remote parts of the network may require the following types of faciltites, eg:

- **loopback**
- **tests**
- **protocol monitoring**
- **switching to bring in redundant components**
- **invoking dial-up from remote locations**

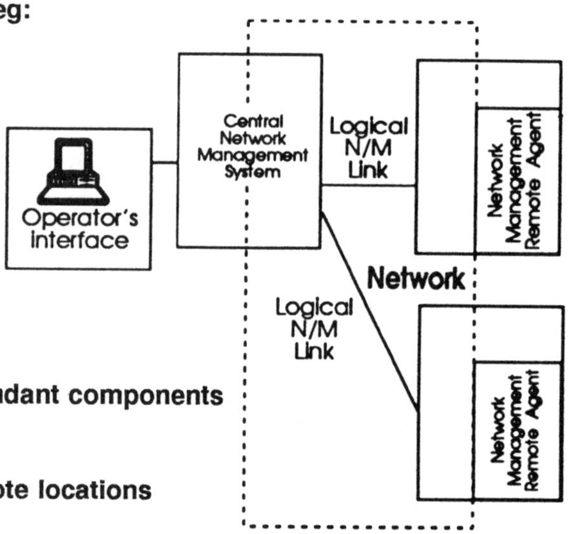

The cost of adding on extra management functions to a given product should be estimated when comparing rival products

Figure 3 Managing the Physical Network (Fault Finding)

In a large number of commercial organisations there will exist several physically discrete server systems. Individual users may not need to access more than one server systems, but there may be several server/user communities that could share the same network resource to derive economies of scale. If there is also a single electronic mail system for the organisation, then the requirement for switching and multiplexing functions can make a convincing case for packet switching (*see* Figure 4).

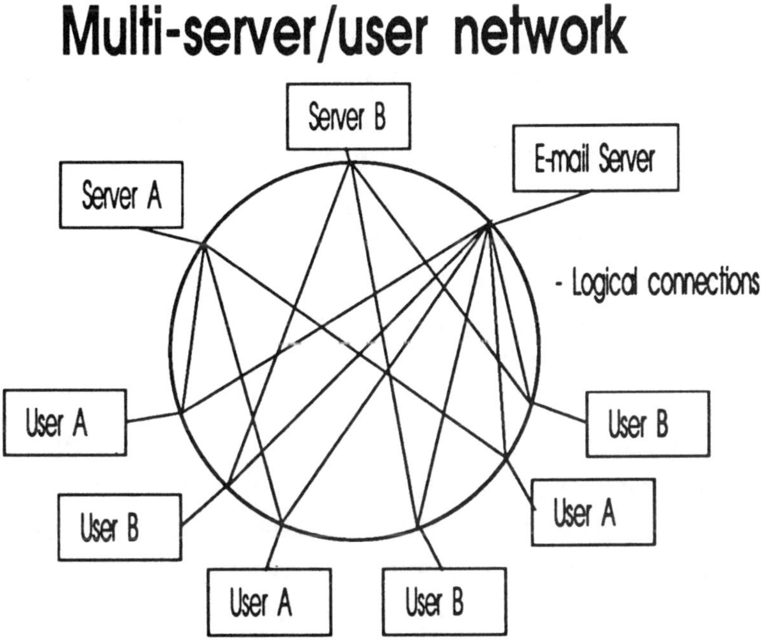

Figure 4 Multi-Server/User Network

1.3 Public and private networks

The choice between public or private networking can involve a number of issues, such as cost, policy (for example the organisation may require direct control of critical business resources) and management (for example the problems of acquiring staff with the right specialist skills).

Having chosen a private or public or facilities managed network (or a mixture of the two), the fundamental management problem is the same, which is to provide the level of service required by the users whilst maintaining control over the costs. It may also be necessary to develop systems for analysing network performance and utilisation. With public and facilities managed networks, the costs (after set up) may be only the fees for the service, which should be known at the outset. However, in public networks such as BT's X.25 Dataline service and its Mercury equivalent, there is usually a volume-related charge which may result in an unexpected cost if the actual amount of data sent is greater than predicted. A breakdown of how this data is generated may provide useful information on both current and future requirements as well as budgetary details.

The performance of the service should be analysed; availability can be analysed by recording losses of service and comparing them against the level agreed with the service provider. Other measures of performance, such as response time and loss or corruption of data, should also be analysed. Where an organisation uses a public network, the detailed management may be delegated to the network provider. With a private network, further management is required by the internal service provider. Development and management is discussed in detail in Chapter 2.

1.4 Structure and role of X.25

This section provides a brief description of how X.25 is constructed and how it functions. More detailed descriptions may be found in other works, including, of course, the CCITT recommendation document. The recommendation defines the interface between the data terminal equipment and the packet switched network, but it does not describe the potential benefits of its use, neither does it define how the network functions internally, nor what protocols are used between the switches. The function of packet routeing is accomplished by proprietary methods developed by the manufacturers.

The X.25 recommendation specifies an interface with three levels, which correspond to the bottom three layers of the Open Systems Interconnection (OSI) model:

LEVEL 1 The Physical Level where electrical signals pass between the Data Terminal Equipment (DTE) and the Data Circuit-terminating Equipment (DCE). A DTE would normally be a Terminal or Front-end processor. A DCE would be a modem or digital circuit terminator. The Physical Layer protocols specified for X.25 are CCITT V24, V35, V11.

LEVEL 2 The Link Access Procedures which provide error-protected transportation of packets across the DTE/DCE interface. The recommendation refers to the Packet Switched Network as the DCE, which is a little confusing because in other standards e.g. V24, it is the modem. The reason is because the recommendation was written from the point of view of an interface between a Public Data Nctwork (PDN) and an end user (DTE), where the modems/digital circuit terminators would be regarded as part of the PDN.

The protocol specified is Link Access Protocol B (LAPB), which is also known as High-Level Data Link Control (HDLC), or Asynchronous Balanced Mode (ABM) protocol. There was another HDLC protocol defined for X.25, called LAP, which is no longer commonly used. Beside error protection, the protocol provides flow control and data transparency. This latter function is provided so that end users can send any character/bit pattern in the data stream without the fear that it will be confused with special characters required by the protocol.

LEVEL 3

The Packet Level controls the transfer of packets across the DTE/DCE (where DCE = Network) interface. It multiplexes virtual calls to and from the DTE on to the physical link (level 1) controlled by the Link Access Procedure (level 2). Each virtual call is allocated a Logical Channel which is retained for the duration of the call. The Logical Channel is allocated by the DTE or DCE depending on the call originating direction. Originally, three types of call were specified:

- Datagram
- Permanent Virtual Call (PVC)
- Switched Virtual Call (SVC)

A Datagram is an example of a connectionless protocol, where packets are sent to the network without any requirement for a call set-up or call clear phase. Datagrams were dropped by the time that the 1984 version of the recommendation was published and were not offered by any Public Data Networks. SVCs consist of a Call Set-up phase and a Call Clear phase with, optionally, data packets in between. The option is offered because Call Set-up and Call Clear packets can carry data. A call consisting of a Call Set-up and a Call Clear packet is called a Fast Select.

PVCs are the same as SVCs except that they are set up and cleared at the time when the system is configured. They no longer tend to be offered by PDNs and are rarely used in private networks. A standard method of flow control is used which limits the number of packets sent and unacknowledged, on each channel, to a certain number. The default is two, but is negotiable. There are numerous standard and optional facilities available, the main ones being:

- Interrupt and recovery
- Retransmission at the request of the DTE
- Call redirection
- Call barring
- Closed user groups
- Reverse charging
- Call statistics on cleardown
- Hunt Groups

It should be remembered that not all these facilities will be present in the X.25 services and systems that are available worldwide, because some are optional and there may be little user demand for them. The CCITT recommendations are revised and reissued every four years after a plenary assembly. The last revision was in 1988 with the reissue in the following year. However, it can take a few years before some of the revisions are implemented in products; some 1984 features were not available until four years later. Manufacturers will give priority to implementing facilities, on their networks, that have greatest user demand, e.g. SNA (IBM System Network Architecture) interfaces. In addition, the PDNs do not normally implement new versions of software until some time after they are available. This is because of their desire to maintain stability in their networks.

One of the most common methods of accessing an X.25 network is through a Packet assembler/disassembler (PAD) (*see* Figure 5). This method allows network access from terminal equipment with an asynchronous interface. If the network is a PDN, the cost of this type of access is relatively low compared with X.25 Packet access. Access is governed by the Triple-X standards which are CCITT recommendations X3, X28 and X29. These standards describe the facilities that a PAD provides and the procedures for access, initialisation, exchange of control information and user data between an asynchronous terminal (DTE) and a PAD. There are also details of the procedures for the exchange of control information and user data between:

- one PAD and another;
- a PAD and a Packet Mode DTE.

The limitations are that the standards only allow one virtual call per physical access link and do not provide any inherent error correction. Only the most basic error protection is provided using character parity and data echoing. Another problem is that the UK public services have a maximum speed of access of only 2400 bits/sec.

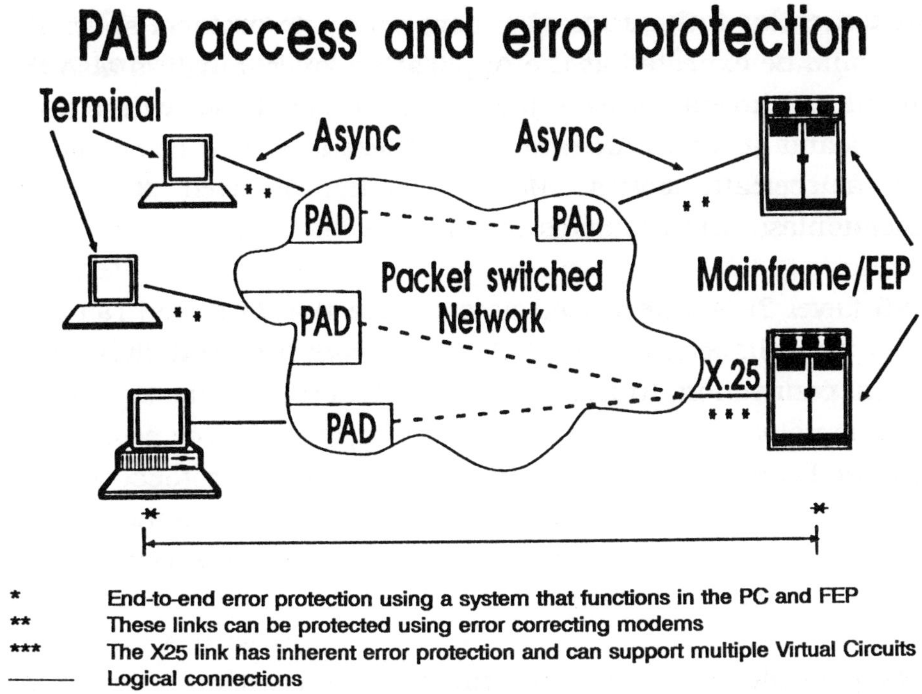

Figure 5 PAD Access and Error Protection

The single Virtual Call (VC) limitation was sufficient when simple asynchronous terminals were used, but now that PCs are being used, multiple VCs would be useful. Error correcting modcms can be used on the access link and this facility is now being offered on the PDNs. With PCs, systems for error correction between end user systems are available, the functions residing in the PC at one end and the FEP, or equipment attached to the FEP, at the other.

1.5 Communications standards

Lower layers

The three levels of X.25 can be categorised as particular implementations of the bottom three layers of the OSI model.

Within each layer there may be more than one protocol specified. This would be expected at the Application layer, but it is also the case at lower layers, for example at the Physical layer, where there are V24 and several LAN protocols. There is a further categorisation at certain layers into connection-oriented (CO) and connectionless (CL) protocols (*see* Figure 6).

X.25 (level 3) is categorised as a Connection Oriented Network Service (CONS), because the Call set up procedure initially sets up a logical connection, in this case to a DTE on the sub-network (ie the same network). It also provides the means of error protection, using the Data Link layer. A connection-oriented protocol may be required where some details of the protocol have to be negotiated dynamically, for example maximum packet size, flow control and quality of service.

It is possible, of course, to transmit data without this prior co-ordination. Ethernet and Token Ring Local Area Networks, which have to provide fast transmission and have low error rates, tend to use a connectionless network service (CLNS). A well-established example of a CLNS protocol is the Internet Protocol, which is the IP of TCP/IP. TCP stands for Transmission Control Protocol. It is part of a suite of protocols promulgated by the US Department of Defence, originally to simplify their systems interconnection and procurement. The protocols are now widespread in the commercial environment. CLNS tend to rely on a connection-oriented transport protocol (for example TCP or OSI Transport Layer Class 4) to provide error protection and quality of service, etc.

Connecting two end users may involve using one or more LANs and X.25 public and private networks. The users of the Network Service should be unaware of the underlying transport media, only requesting a certain quality of service from it. One of the most important functions of the Network Layer is relaying and routeing

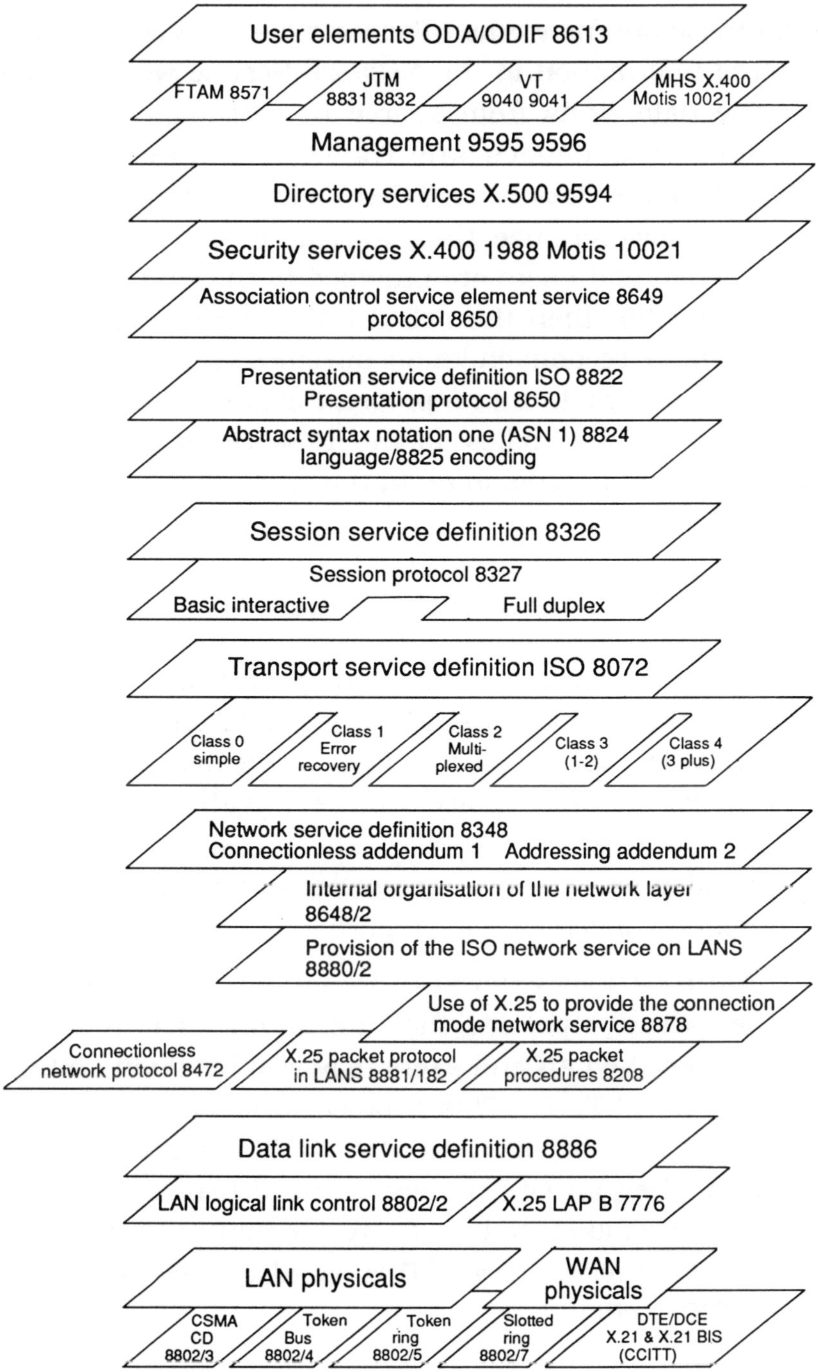

Figure 6 OSI Protocol Stack-key Standards

through these sub-networks. The user has to provide destination addressing information to the Network Service which will enable it to route data units from source to destination. In the OSI model, these destinations are Transport Entities identified by Network Service Access Point (NSAP) addresses. Particular types of systems, however, use their own addressing standards. For instance, when end-users are attached to the same sub-network, for example X.25, then the called DTE address would be used. The 1984 X.25 recommendation specifies a field for the NSAP address in the packet, but no processing is specified for it. NSAP relaying and routeing relies on the 1984 recommendation being implemented and functionality, in the relaying networks, to route using the NSAP address.

Examples of other methods in use are:

- Address translation between public and private X.25 networks. This will tend to be based on the sub-address digits (ie the last two digits) of the DTE address. The translation takes place in the private network or in a gateway between the two. The translation could also be provided between two private networks. This is a standard method conversion (*see* Figure 7).
- LAN to LAN via gateways through an X.25 network, where the gateways provide the translation. The protocol used between the LAN end users may well be a non-OSI protocol such as TCP/IP. This protocol unit is put into an X.25 packet to transport it over the X.25 network (*see* Figure 8). Even when an OSI protocol is used between the LAN end users, it is likely to be a CL type entailing, in this example, CL/CO interworking. This is discussed further in the GOSIP section.

Availability and widespread use of products that support relaying and routeing using NSAP addressing are still in the future.

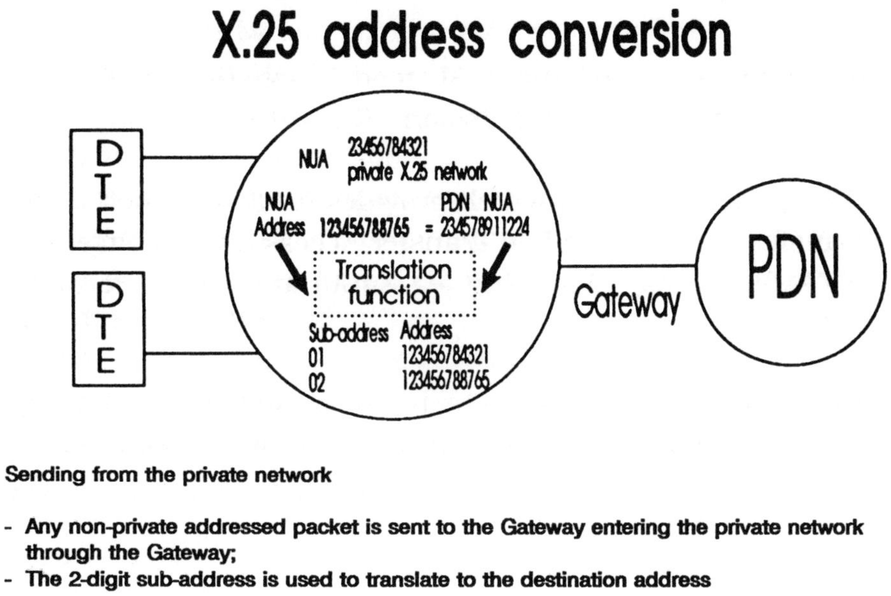

Figure 7 X.25 Address Conversion

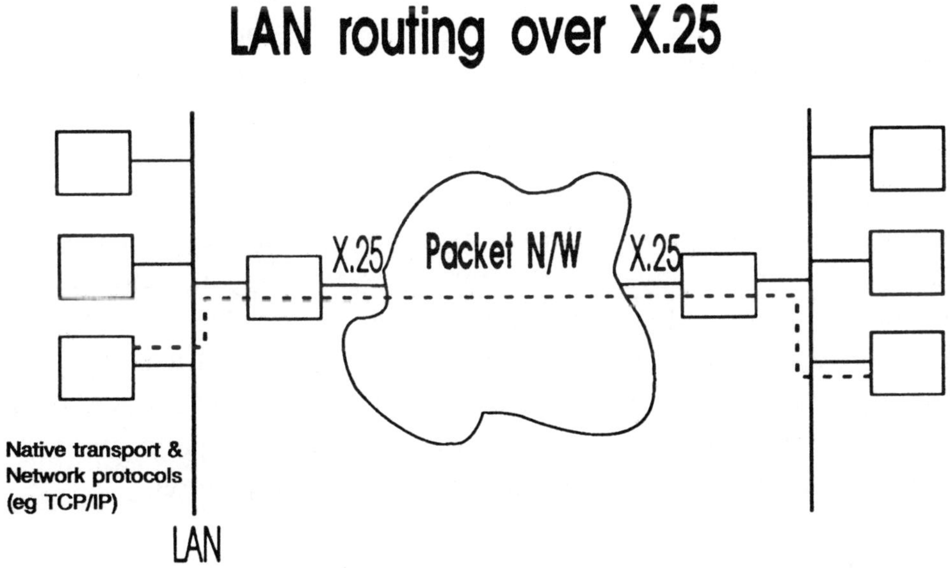

Figure 8 LAN Routing over X.25

Higher layers

The top three layers of the OSI model are concerned with applications rather than transmission. They are:

- the Application Layer which provides functions such as file transfer (for example File Transfer Access and Management (FTAM)), message handling (eg X400) and virtual terminal (VT);

- the Presentation Layer which converts between specific terminal characteristics and generic ones used by an application, for example data translation, formatting and syntax modification;

- the Session Layer which controls the dialogue between applications and recovery through a checkpointing technique.

Transport layer

In the middle of the OSI model is the Transport Layer. Its importance to reliable communications lies in it providing transport connections between user end points (*see* Figure 9). This is the lowest layer in the model containing an end to end protocol. It provides five classes (0-4) of service designed to function with underlying networks of different qualities. Class 4 provides the most functions, being able to detect and recover from errors. Most importantly, it can provide resilience against network failure by making use of multiple network connections.

The quality of service of the connection can be monitored with reference to four parameters: availability, throughput, delay and residual error criteria. The transport service described is connection oriented. This is by far the most commonly used type of transport service, enabling protocol parameters to be negotiated when setting up the connection (for example class, data unit maximum size, delay, residual error rate). TCP, the transport layer of TCP/IP from the Internet protocol suite, also provides a connection oriented service.

The Packet Switched Systems described in Chapter 4 provide transport layer type functions (not OSI) transparently to users, which means that if the network is configured with resilient routes, the network will re-route in the event of a failure.

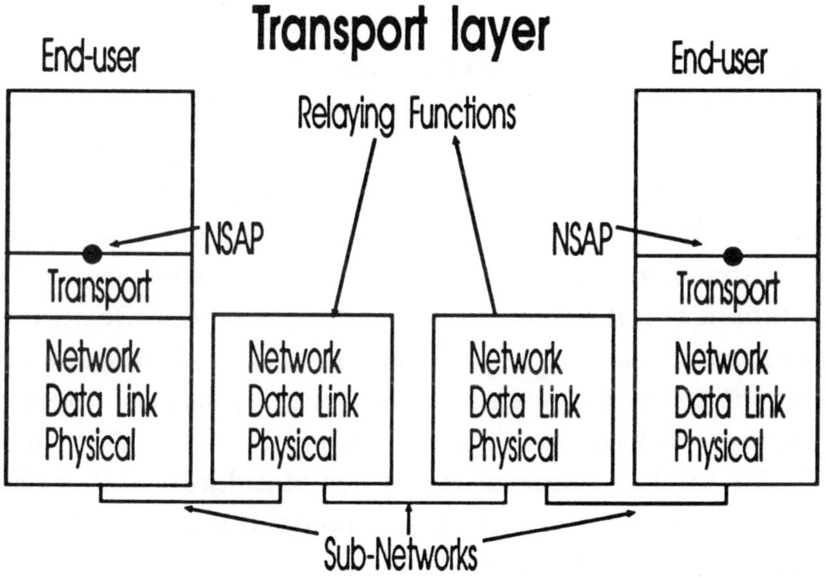

Figure 9 Transport Layer

1.6 GOSIP (Government OSI Profile)

Purpose

The UK Government OSI Profile is a procurement specification covering OSI based data communications for UK Government civil administration. Its stated aims are:

- to facilitate procurement and acceptance testing of communications-based products
- to ensure that different and separately procured departmental systems can interwork to an assured level of functionality
- to provide a clear specification to manufacturers on which to base strategic product development.

GOSIP is guided in its preparation by the Central Computer and Telecommunications Agency (CCTA), the IT advisory body to the government. At a simple level, GOSIP has defined a sub-set of the available standards that suit the user environment, ie the UK Government IT services (*see* Figure 10). The present and likely availability of OSI compliant products has also been taken into account, so that it will be of value to users in the short term. The UK Government is not alone in producing this type of OSI profile; many governments and other large IT users have done the same. In addition, there is European Community legislation: Directive 87/95 requires OSI conformance in all public sector IT tenders over 100,000 ECUs. Some of the other standards and profiles with which GOSIP will attempt to be compatible, assuming UK Government requirements are met, are:

- other government profiles, for example US GOSIP;
- British Telecom's Open Network Architecture (ONA);
- General Motors Manufacturing Automation Protocol (MAP);

- Technical and Office Protocol (TOP) from Boeing;
- US National Bureau of Standards (NBS) Implementation Agreements;
- US Corporation for Open Systems which are intended for the development of testing services (*see* Figure 11).

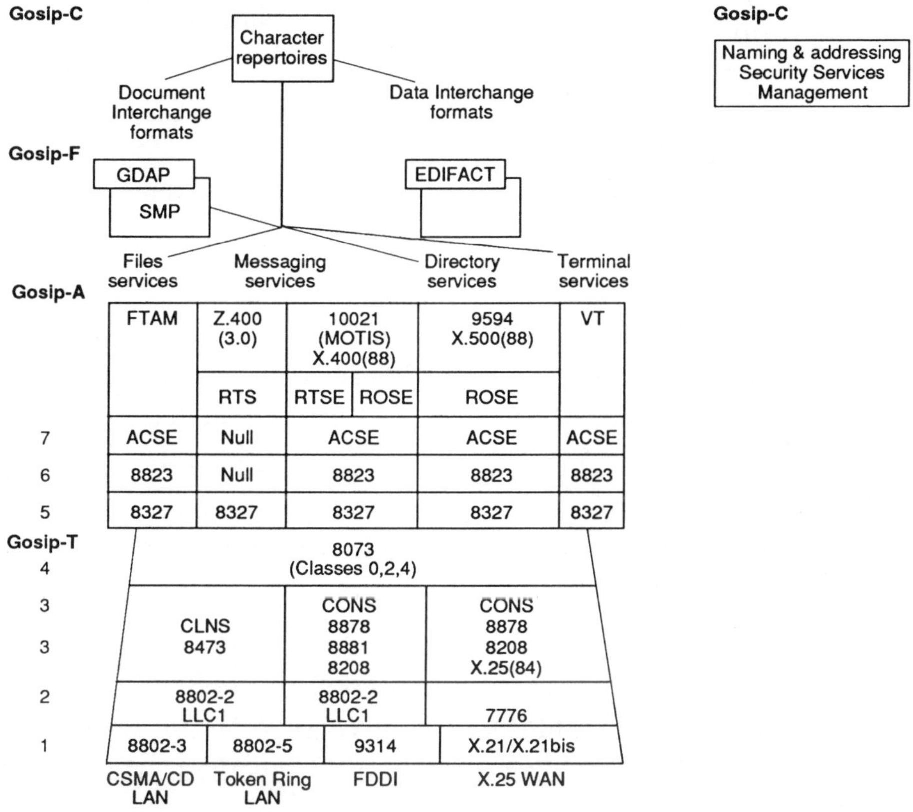

Figure 10 UK GOSIP Stack

The GOSIP profile defines two levels of testing:

- conformance testing to determine that a product will meet the requirement in the standard to which it is claimed to conform;
- inter-operability testing to determine that the product will interwork with other systems as intended.

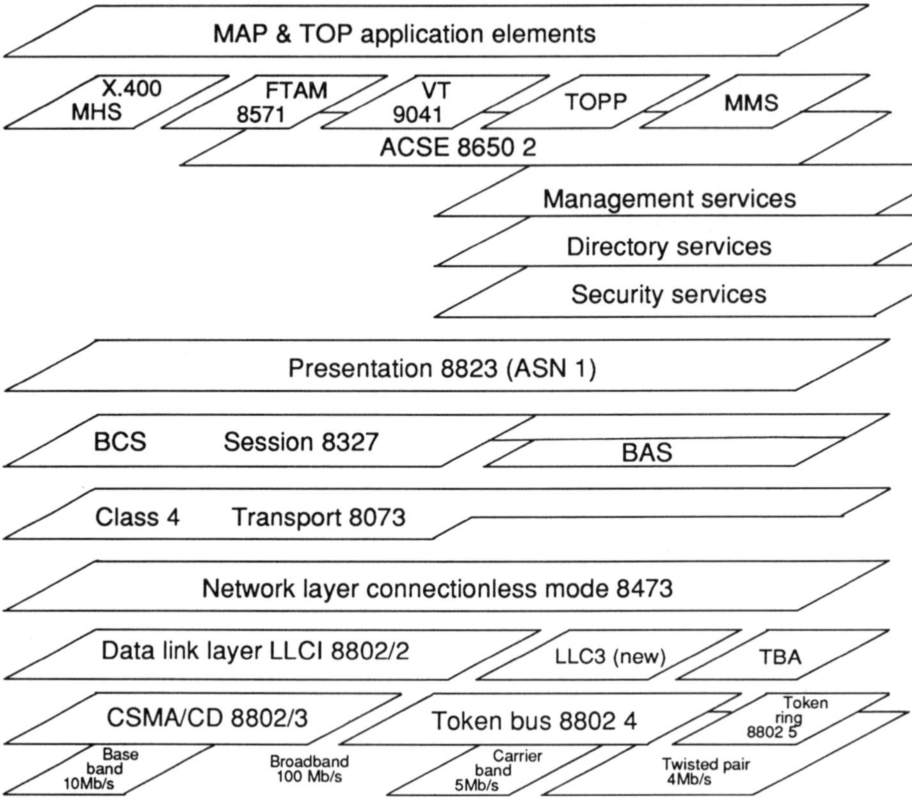

Figure 11 MAP/TOP Profiles

For most sub-profiles (ie standards), GOSIP tabulates the selections and constraints required for conformance. In the actual procurement situation, additional user requirements, beyond the minimal conformance requirements, can be specified. This should fulfil the objectives of:

- a basis for contractual acceptance;
- providing a test specification;
- identifying the level of interworking that should be achieved with other systems.

GOSIP expects conformance testing to be part of a product's normal development cycle rather than part of the procurement process. As this type of testing should be undertaken by a developer anyway, the additional requirement should be simply for properly documented testing details and results. In addition, there is a requirement to submit product details to a testing service, assuming a suitable one is available. The availability of suitable testing services will be monitored by CCTA and users and suppliers kept informed about them. CCTA will also encourage the development of internationally recognised conformance test suites. The advantages to the user, in terms of time and cost, are clear. There are similar advantages to suppliers as they should not have to undertake conformance tests each time the product is bid to a different user.

The situation with interoperability testing is different in that covering all combinations of product interconnections is impractical, so this aspect of testing may still have to be undertaken during acceptance testing.

Connectionless/connection-oriented network services

It can be seen from the diagram of the GOSIP standards that both Connectionless (CL) and Connection-oriented (CO) Network Services are embraced, Connectionless catering for the overwhelming number of LAN products. In earlier releases of the GOSIP profile, the CL standards were described as interim and migratory. This description has disappeared from the latest version (3.1) and reflects the product situation as it presently exists. The recognition of CL protocols also aligns UK GOSIP with the US GOSIP, MAP and TOP profiles. GOSIP also recognises the requirement to interconnect CL and CO networks, e.g. LAN to X.25 to provide a wide area network router. The profile specifies use of the CO/CL Interworking Unit (IWU). It is not standardised, although the principles of its architecture are documented and in the public domain. The CO/CL IWU, also described as a Distributed Systems Gateway (DSG), is covered in the ECMA-TR21 document (available from ECMA) which specifies a mapping and converting function at the Transport layer. This means that there are two OSI layer 1-4 protocol stacks.

The IWU will have to deal with problems such as addressing, quality of service using the Class parameter, flow control and error recovery. How effectively these problems are dealt with may, of course, have an impact on system reliability and performance.

It should be noted that, in UK GOSIP and the other profiles, the Transport Layer standard used is Connection-oriented. In GOSIP it is IS8073, the ISO Connection-oriented Transport Layer standard. GOSIP also specifies the ISO Connection-oriented standards for Session and Presentation Layers.

Frame relay and fast packet switching

Frame Relay and Fast Packet Switching technology are based on related CCITT recommendations for ISDN services. Frame Relay is being implemented for data only using variable length packets, whereas Fast Packet is implemented for short, fixed length packets which can contain voice, video and data. For Frame Relay, a number of manufacturers are planning to provide implementations in existing X.25 networking systems. For this data only technology, the functionality, specified in the recommendation, already exists in networks. The frame format will have to be mapped to the functions. For Fast Packet, there is at least one system already available.

Frame relay

Frame Relay can provide a simplified technique for the transmission of data packets compared with X.25. It is a standard protocol, using the CCITT recommendation I122, which specifies relaying and routeing functions at the Data Link level (ie layer 2 of the OSI model). The purpose is to specify functions for networks to transport packets at far higher speeds than with existing X.25 networks; up to 45 Mbits/sec is being tested. Frame Relay is an ISDN (Integrated Services Digital Network) packet mode service in which signalling (ie Control (C) plane) and data (ie User (U) plane) are logically separate.

Communication over a frame relay network requires a virtual path to be established before data is transferred (*see* Figure 12). This may be done at configuration time (permanent virtual calls) or at call set up (switched virtual calls) by the end user device. Call control signalling between the end user device and the network takes place via the ISDN D (ie call control) channel using procedures defined in CCITT recommendation Q931 and between nodes using the ISDN system. Sending data requires frame relay interface capability on the ISDN B channel (64kbs) plus V120 functionality if segmentation and re-assembly is needed. However, frame relay terminal adaptors should be available for end user devices without ISDN compatibility.

An immediate adaption will be for the interconnection of LANs and workstations where there is a requirement for switching flexibility and large amounts of bandwidth. Other adaptions/protocol conversions will be available for other end user types (for example X.25, Async).

The Frame Relay recommendation describes several services which vary in their richness of functions, including a service which provides a much simpler set of functions than X.25.

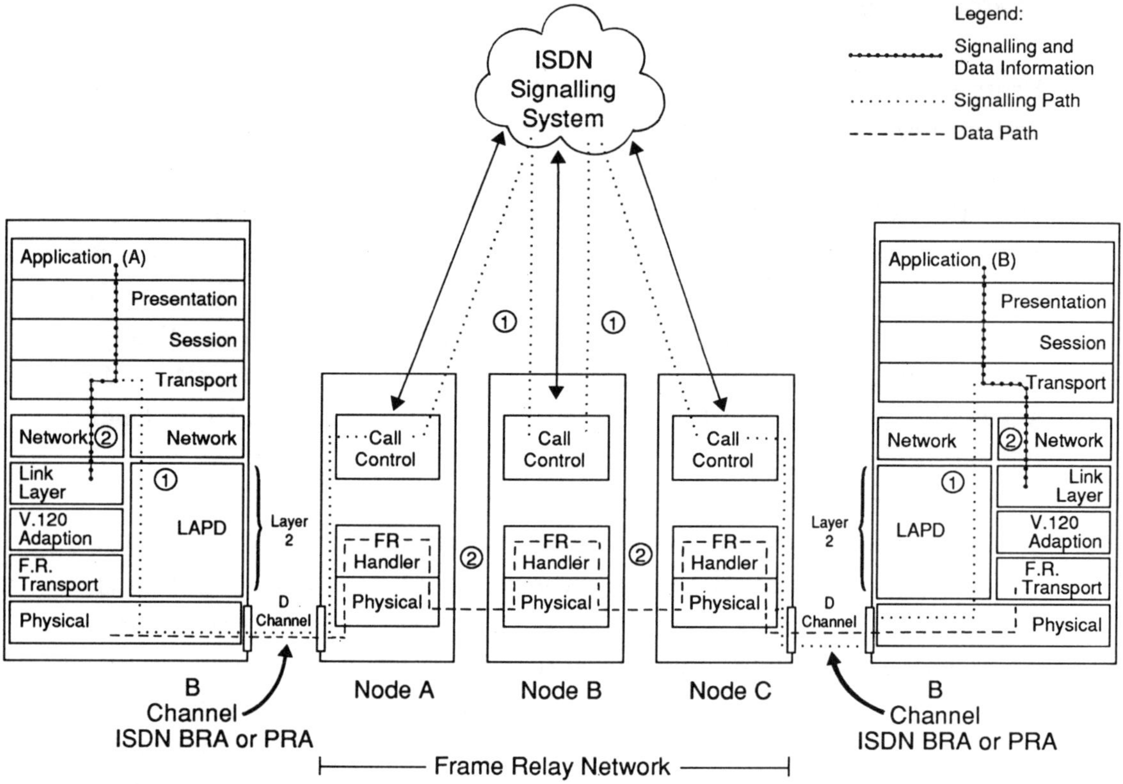

Figure 12 Frame Relay

Significantly, there is no flow control nor error control in this service. These functions are left to be performed in the end user systems higher layer (for example Transport) protocols. There is an underlying assumption that the networks being employed will introduce few residual errors. Any error detected will result in the frame being discarded by the network. Frame Relay is required to deliver packets in the order that they are sent, and this should be achieved implicitly through the connection oriented virtual path routeing.

A number of packet switch suppliers state that they have firm plans to provide Frame Relay on their existing equipment. Physical interface speed should not prove a problem, as existing interfaces are capable of high speeds. The functions specified in the CCITT recommendation already exist in hardware and software. The networks, however, will also have to map the functions to the Frame Relay packet format (*see* Figure 13).

Frame relay packet format

a) Format of Frame Relay Header (FRH)

<table>
<tr><td colspan="6">DLCI (higher order)</td><td>C/R</td><td>EA0</td><td>byte #1</td></tr>
<tr><td colspan="4">DLCI (lower order)</td><td>CF</td><td>CB</td><td>DE</td><td>EA1</td><td>byte #2</td></tr>
<tr><td>8</td><td>7</td><td>6</td><td>5</td><td>4</td><td>3</td><td>2</td><td>1</td><td></td></tr>
</table>

DE - Discard Eligibility Bit
DLCI - Data Link Connection Identification
CF - Congestion Forward Notification
CB - Congestion Backward Notification
C/R - Command/Response

b) Format of V.120 Terminal Adaption Header

E	BR	res	res	C2	C1	B	F
8	7	6	5	4	3	2	1

E - Extension Bit
BR - Break/HDLC Idle Bit
C1 - Error Control Bit 1
C2 - Error Control Bit 2
B - Segmentation Bit B
F - Segmentation Bit F
res - Reserved for Future

c) Format of V.120 Control State Information Field

E	DR	SR	SR	RR	res	res	res
8	7	6	5	4	3	2	1

E - Extension Bit
DR - Data Ready Bit
SR - Send Ready Bit
RR - Receive Ready Bit
res - Reserved for Future

Figure 13 Frame Relay Packet Format

Fast packet

The function of Fast Packet is broadly similar to Frame Relay except that voice and video are also handled and short, fixed length packets are used (*see* Figure 14). In the future, it is intended that Fast Packet networks should handle interface speeds of hundreds of megabits. The initial target market for Fast Packet Switches is that occupied by the high speed Time Division Multiplexors (TDMs). Fast Packet technology enables better use to be made of the bandwidth than with TDMs. TDMs have fixed channels to which dedicated bandwidth is allocated, for example 30 x 64k channels on E1 (such as BT's ISDN30) 2 megabit circuits. Fast Packet allocation of bandwidth is dynamic, being based on user requirements at a given moment in time (ie like packet switching). The technology should be especially attractive to users who do not already have any great investment in high speed TDMs.

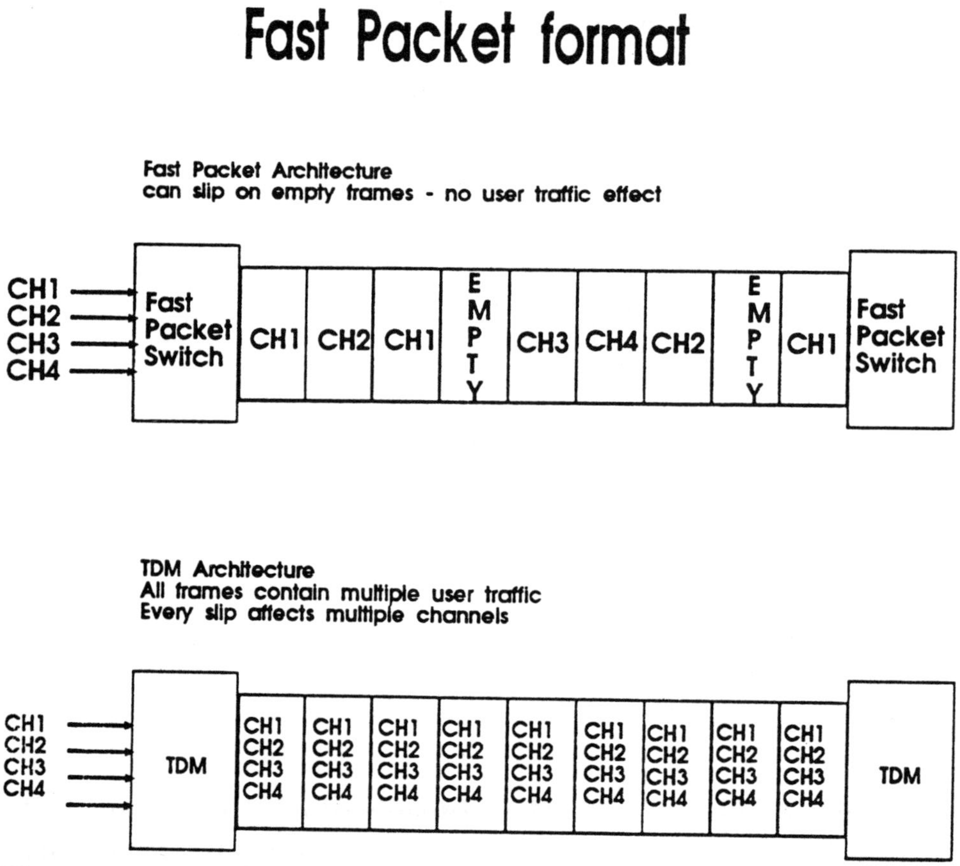

Figure 14 Fast Packet Format

2

Development/management of telecommunications networks

2.1 Background

Reliable telecommunications have been extremely important to business systems for many years. In earlier years, a network was usually dedicated to a particular on-line application, for example branch banking and order entry. The networks caused some development problems but very few compared with the problems of developing on-line applications. The requirement in organisations for more on-line applications has often been satisfied by the introduction of a separate network, but as the networking requirements increased, the costs inevitably spiralled, and planners have looked to technology to reduce these costs by the means of sharing networks between applications.

Organisations have become more and more dependent on applications and networks which have high availability, communicate accurately and process users' transactions within a few seconds. LANs are used to share resource and wide area networks to interconnect them. More complex technologies (for example Packet Switching, LANs), increasing size and connectivity requirements have inevitably made the development and management of networks more difficult. Designing multi-user networks with high levels of service has added an extra dimension to the difficulty.

Reliable networks are fundamentally important to the success of a large number of businesses. The loss of service through host or network failure, for instance in the airline business or in some financial applications, even for a few minutes, can result in losses of tens to hundreds of thousands of pounds. Delay in the implementation of applications has also been the cause of losses that have run into millions of pounds. These extremely long delays are generally due to problems with the development of application software. Implementation delays caused by communications problems are often due to poor preparation, particularly in the specification of requirements and interfaces, as well as inadequate testing and not training sufficient staff in the necessary network management skills.

Successful development and management of networks is therefore critical. The success must be quantifiable, the most important measure being the performance of the service to the user, compared with demand. Service Level Agreements should be negotiated between the supplier and the user which detail the contracted service. The performance of the network should then be analysed, which requires putting systems in place to record faults and other problems as well as to analyse performance. Service availability is dramatically affected by the speed of detection of faults and service restoration, so these problems must be also addressed. In the longer term, the means to detect problems such as bottlenecks are required, as these will cause response times to increase. Information about both current traffic and future expansion requirements is also important so that smooth and flexible network growth can be planned and costed. Organising these items requires a structured approach:

- procedures to describe the work that is carried out and who is responsible for the component parts of it;

- management systems, which may be bought off the shelf (for example Protocol Monitors) or specially developed (for example Performance Analysis), to process information relevant to managing the network;
- staff with clearly defined responsibilities, who are trained and skilled in the tasks they are required to undertake.

Evolving and maintaining this structure should be quality assured by reference to a quality standard, for example BS5750 (a quality standard in widespread use).

2.2 Summary

Successful network management entails the organisation of a number of activities that provide information on which long and short term network decisions can be made. The activities are generally split between different areas of responsibility. An example of a responsibility structure is as follows:

Operations whose general responsibilities are:

- diagnosing network problems (for example line faults);
- implementing network recovery (for example a dial up line);
- carrying out network changes (for example configuration changes);
- providing information on faults and other network events (threshold exceeded).

These activities tend to be reactive and have a real time effect on network performance. Their documentation responsibility is to produce raw network event information for analysis.

Administration whose responsibilities are:

- managing the network inventory;

- managing fault records;

- managing network utilisation records;

- managing network changes;

- producing and network performance figures and reviewing them with users and suppliers;

- producing utilisation figures;

- providing network running costs.

These activities cover day to day record maintenance and analysis. They will also monitor the effectiveness of the systems and equipment used by Operations.

Technical Support and Analysis, whose responsibilities are:

- analysis of trends in service levels and network utilisation;

- network technical problems;

- assessing and reporting on remedies.

This is a role that tends to be poorly defined and consequently neglected in many organisations. One problem is that the information required is not easily gathered and systems may have to be set up to obtain it. The trends in performance and utilisation provide important information for future plans. Dealing with technical problems which fall outside the brief of Operations and

Administration should complete the allocation of responsibility for addressing network performance problems.

Network Planning whose responsibilities are:

- requirements analysis (both long and short term);
- assessment of network alternatives;
- network design and implementation planning;
- project management.

These activities are concerned with the development of networking to deal with demand. They have to plan for capacity to be available when it is required. They must provide information from which costs of development and running can be estimated. They should also be able to quantify the benefits of the service to users and the organisation as a whole. This analysis should cover present and potential costs versus benefits.

There is also the need to discuss activities for network management following a route from initial requirements through planning to implementation and support. Quality Assurance may be considered to be everyone's responsibility, but ultimate responsibility for QA should reside with whoever is in overall charge of providing the communications network.

2.3 Assessment of business requirement and data collection

Background

It is fundamental to success that the business requirement is clearly understood. The Business Analyst must uncover all necessary information to provide a level of telecommunications service that fulfils the requirement. The Network Planner will then use the information to prepare a proposed implementation plan.

Business analysis

Information requirement

- analysis of the business requirement should produce the following information.

Locations of user equipment

- implementation schedule;
- response time (to establish the slice for which the network is responsible);
- required network availability for the user equipment connected at each location;
- traffic profile:
  - volumes,
  - message sizes,
  - peaks,
  - routes (for example terminals to mainframe, all terminals interfacing to each other);
- interfaces available on the users' equipment.

Ideally, the Business Analyst should be aware of all impending, network related, developments well in advance so that the Network Planner can do his Capacity Planning job. It is part of the Analyst's job to be aware of this information and to make the organisation aware of the advantages of providing the information. These advantages are essentially resource optimisation, cost optimisation and shorter lead times in providing service to users.

Network and transaction processing trade-offs

The application design and the type of terminal equipment chosen can effect the network requirements and costs. The Business Analyst should be party to these decisions as there are certain trade-offs to be considered. For instance, choosing limited function terminals (for example VT100, DRS20, IBM3270) requiring an on-line host has the following effects on network and transaction processing requirements:

- all transactions have to be sent over the network;

- all screen and print formatting characters have to be organised in the host and sent over the network;

- high availability of host and network is required as the user is dependent on the host for the application service.

Choosing intelligent terminals with local storage can have the following effects:

- the ability to process some transactions locally;

- no need to send formatting data with transaction as this can be done locally;

- non-interactive data (for example, files) can be sent to local storage during times of low network activity and do not need a user in attendance;

- lower network availability required as users are less dependent on a host;

- the application development is more complex;

- end user equipment is more expensive and needs to be very reliable.

Additionally, if the application demands a packet switched network and the terminal equipment does not have inherent X.25 support, then there will be additional expense incurred for the provision and management of protocol conversion.

2.4 Establishing the network requirement

Traffic Profiles

Routeing

The business requirement and application design will have determined the data routeing that the network has to support, whether *hierarchical*, with all users connected to a central host, or *flat*, with all users requiring connections to each other. Both types of routeing may be required in organisations with central and regional offices where data is routed between regional offices as well as from regional offices to the central office. In this case, there may be a requirement for packet switching, while LANs may be required in the offices to support reasonably large populations of terminals/PCs. It should be remembered that in many organisations there may be other applications, on other hardware, to which users require to be connected. All requirements have to be analysed and opportunities for network sharing assessed.

Message sizes

These will have been established in the application design and are significant for the calculation of response time because the mode of transmission is serial and because equipment buffer sizes have to be estimated.

Volumes and peaks

Transaction volumes for new interactive systems are notoriously difficult to estimate. In an interactive banking system, for example, the number of transactions can be double that in a batch system because although the number of debits and credits remain about the same, on-line systems provide enquiry transactions and an ease of access to information that is absent in batch systems. It should be possible to make an estimate of volumes based on the functions available in the system combined with discussions and tests with the users. Another possibility is to find an organisation with a similar existing system. Of course, this assumes that the other organisation is monitoring the traffic. It should be possible to predict the occurrence of peaks with some confidence if daily, weekly, monthly work schedules, based on new system functions, can be estimated.

Locations and take-on schedule

These two items go hand in hand. The planner will need to know where there are opportunities for sharing resource, e.g. by routeing to the same switch, sharing the same PAD/Concentrator or the same multipoint line, and when he has to commit resource to implementation. How much change is likely to occur, i.e. the frequency of changing and adding locations, will also be an important factor.

Availability

Availability is the percentage of time that the service is accessible by the user during the total time that service should be accessible. Individual component failure is of no interest to the user, only access to the service when he wants it. However, simply expressing it as a percentage may not be sufficient: for example, 95% would suggest service unavailability of one working day each month. If the maximum down time for a location cannot be more than 4 hours and this cannot occur more than twice a month, then the design must take into account these factors.

Response Time

Response time is very important to the customer, and is generally regarded as the elapsed time from pressing the enter key to receiving the response at the terminal. The Analyst needs to determine the amount of this time that is taken up in hosts and other processing equipment and thus calculate what remains for the network.

Interfacing

Responsibility will need to be taken for establishing the OSI standards within an organisation. For practical purposes, the higher layer interfacing choice may have already been made through the choice of end user systems. A 'higher layer' task that will often have to be addressed by the Network Planner will be Presentation Protocol conversion (e.g. VT100 to IBM3270 format). This is usually when the conversion has to be accomplished in the network and management of the converters is the responsibility of the Telecomms Group.

It is important to determine what Transport Layer functionality the end user systems provide, for example:

- error detection;
- error recovery;
- flow control.

Ideally, it should provide end-to-end resilience, given that there are alternative connections into the network. Otherwise, the network may provide this functionality, perhaps by X.25 Call Redirection or Hunt Groups. Most packet switching systems provide Transport Layer functions, within the network, using proprietary protocols. The effect of this is that, between any two network end points, packets will be re-routed transparently to the user, in the event of one route failing and another being available. Where the data is likely to be routed through more than one network system (ie sub-system), planners must ensure that routeing information (ie addresses) can be properly handled.

Attention should be paid to the structure of addresses as this can significantly affect the routeing efficiency in some suppliers systems (for example some packet switches).

The requirements should be defined in detail as the Data Link layer flow control and error correction may be able to provide some of the functions.

2.5 Network planning

General Responsibilities

The general responsibilities of the Network Planner are:

- assessment of future network requirements from present and new user applications;

- producing plans for networking enhancements by:

    - optimising present capacity,
    - implementing additional capacity,
    - introducing alternative networks.

- providing estimates of future costs for:

    - budgets,
    - reviewing tariffs.

N.B. It is vital that account is taken of Network Management requirements and resources.

Project Responsibilities

For specific user requirements, the responsibilities of the Network Planner are:

- planning how the requirement can be implemented in terms of:

    - using existing capacity,
    - additional capacity required,
    - interfacing network to host/terminals,
    - testing,
    - scheduling the implementation,
    - managing the service to be provided;

- providing a level of service with the required:

    - availability,
    - response time,
    - data delivery integrity;

- estimating development and management manpower and skills;

- estimating costs of development and management and a proposal for cost recovery (for example future tariff increases);

- organising management of the project with co-operation and committed help from:

    - the business analyst,
    - the user,
    - host/terminal system developers and suppliers,
    - network equipment and media suppliers,
    - network operations and administration staff,
    - senior management from the service provider and user to whom progress versus plan and cost versus budget is regularly reported.

The development of large projects can be chaotic and an organisation must be prepared for its normal running to be disrupted. Projects must have clear, time-related objectives, punctuated by milestones, with progress being measured by their achievement.

2.6 Network design

If a corporate, multi-purpose network already exists in an organisation, it may be possible to extend its use to the new requirement. If nothing suitable exists, the business analysis may have already made some design decisions obvious. The following describe the more important factors that have to be considered in the design. Given the information from above, the Planner must produce a design to fit the requirements. Where this is difficult or where costs appear to be high, he must negotiate a solution with the User, Application Developer or both.

Availability

Availability can be defined as :-

$$\frac{\textbf{MTBF}}{\textbf{MTBF+MTTD+MTTR+MTTRe}}$$

where

MTBF is mean time between failures

MTTD is mean time to diagnose

MTTR is mean time to repair or invoke redundancy

MTTRe is mean time to replace a redundant component with a repaired one.

MTBF should be very high compared the other three factors, and should be fixed, as it is a function of the MTBFs of the network components used in the design. It should thus be predicted with accuracy. The other three are functions of the Network Management systems employed, the skills of the staff, good procedures/working practices and the level of support purchased from the supplier. Without redundancy, there is a practical limit to the amount by which these factors can be reduced. Significant improvements in availability will only be brought about by introducing more redundant components.

Redundancy and Contingency Planning

As redundancy is costly, it must be effective, and the decision to provide it must be made on the cost of the consequences of failure balanced against the cost of redundancy. For instance, the loss of a mainframe for a sustained period, due to fire or flood for example, would result in unacceptable financial losses for most organisations. Downtime of mainframe, Front End Processor (FEP) or major network components, even for a few minutes, can result in losses running into tens of thousands of pounds. Consequently, to provide the required availability, redundant computer centres, mainframes, media and FEPs may be configured. In the networks the following may be necessary:

- links to redundant computer centres, mainframes, etc;
- redundant network management centre/system;
- major trunks alternatively routed down separated physical paths;
- redundant major components, for example switches, multiplexers;
- dial-up from the end-user terminals (for leased-line failure);

- equipment component redundancy (power supply, memory, processor, port);
- network designed so that single failures are not catastrophic.

The network requirements are certain to change in time: volumes may increase, office locations may vary, message sizes increase and new applications will be planned. Account should be taken of this in the design plan in terms of extra capacity, flexibility for growth, manageability and maintainability. It is advisable to build extra capacity into the initial design. A year's estimated growth would not be unreasonable plus an allowance for unforseen growth. In an evaluation it is a good idea to determine the effect on the network of double and quadruple the requirements. This will provide a measure of growth flexibility and cost of growth. The initial contingency should avoid the problems of making frequent network changes, such as downtime, erosion of the provider's credibility and any lack of resources being exposed.

Response time

The network slice is the time spent queueing for the network and the time spent using the service. In a packet switched network, queuing will include time spent in the switches waiting to use a line. As with availability, response times can be reduced at a cost. In the mainframe/FEP, the time is taken up by CPU and I/O service and waiting times. Assuming no major bottlenecks, faster CPU and I/O channel changes will not significantly reduce response time. Similarly, over a packet switched network interconnected by 64kbps lines, there can be little improvement if capacity has been properly planned. It is at the terminal end of the network that the greatest reductions are to be made by increasing the speed of low speed lines to the terminal.

In practice, terminal utilisation on the line into a network will be very low. A teller on the counter of a bank, for example, may serve 30 customers during a busy hour taking about one and half minutes to serve each customer. Apart from a few seconds, the time to serve the customer is taken up by dialogue, teller thinking time and data entry time. An autoteller may reduce the service time by more than a half but still not greatly increase the line utilisation. Even where data is being keyed in to a terminal from a batch of forms and there is full screen output, utilisation is not great. Generally, utilisation of a line will tend to be increased by sharing it between a number of terminals, using PADs, contention devices, multipoint lines and polling and statistical multiplexors. Where message sizes in both directions are relatively small, say a few hundred characters, line speeds may still be kept down to 2400bps. A larger number of terminals on the line may entail raising the speed.

The reason for raising the speed is more likely to be for giving a low response time to transactions with a large message size, e.g. for full screen output.

Statement of requirements and evaluation

Having gathered all the information discussed in the preceding sections a Statement Of Requirements (SOR) can be produced from which suppliers can propose network solutions. The Proposals can be marked against pre-set criteria weighted by relative importance.

2.7 Testing and Test Networks

The major equipment components of a network are controlled by software. Like any software, it is subject to periodic new releases.

All new software releases should be tested before they are implemented on a live network. To do anything else is to take a risk which could cause downtime, resulting in financial loss as well as loss of credibility with the user. This holds true for modems, multiplexers, LANs, network management systems and particularly packet switches.

In a corporate network, supporting user applications for 24 hours a day, 7 days a week, their are clearly limited opportunities for testing new software. Ideally, a Test Network should be used for testing new releases. The cost of the extra equipment should be balanced against the consequences of network failure.

The motivations for moving to a new release are:

- new features and bug fixes which are required to support user and network management requirements;
- to be adequately supported by the supplier, for a reasonable period of time. (See also Migration Planning.)

The aim of testing is to build confidence in the stability of the release. The objectives of testing are:

- to ensure no degradation of functionality compared with the current release. This should include making sure that interworking is still possible with other systems;
- test any new functions and bug fixes that are planned to be used during the time that the release will be live.

Testing/checking should also cover the following:

- audit the manufacturer's own system testing. If this has been performed to high quality standards, then all tests and results should be fully documented;

- audit the release's functional documentation to ensure that it is adequate;

- repairs to 'bugs' from previous releases should be documented along with any known problems in this release;

- there should be available a list of the modules which constitute this release along with the module revision numbers. The revision history should be documented to show what each revision fixed and what testing was carried out;

- a recommended migration path should be available detailing the steps to be taken when a customer moves to a new release.

The evidence of this documentation will be of great help in persuading the customer that the developer has a structure that uses quality standards for developing and releasing systems and software. This documentation system should continue in use during the lifetime of the release, ie as bugs are reported and new module revisions are available. The documentation should be available to the customer.

Volumes are notoriously difficult to set up tests for because of configuring an adequate amount of equipment. Load Generators are another area of difficulty. They do exist for X.25 systems and can be bought or hired along with expertise.

The Test Network can be used for User Testing, particularly in the area of interfacing to the network. If, after this, the user needs networking in his application testing there is no particular reason why he should not use the live network.

Migration Planning

The main types of migration are:

- to upgrade to a new release of software;
- to move users on to a new network;
- to physically reconstruct a network, for example to put an access layer into a packet switched network;
- to support a large, user move, for example moving a computer centre.

The releasing of software was covered in detail in the previous section. For the actual migration, the following should be taken into account:

- all other network related changes should be frozen around the time of the migration. This includes physical changes and changes to configuration tables. If at all possible, users should also be persuaded from making changes. The reason for the freeze is that if a problem arises, the cause can be more easily narrowed down. This is a good general policy to follow for any changes;
- a plan for reverting to the previous release if a problem arises that cannot be solved in the time window scheduled for the migration;
- users should be informed of any possible downtime. Their co-operation should be sought to perform some tests after the migration is completed. All users should be informed when it is over;

- the migration must be properly supported by the supplier and Network Management staff before, during and after it occurs. The after is often neglected but presents the opportunity to monitor the network for any behavioral changes that may be caused by the migration. Being properly supported by the supplier includes provision of the documentation listed in the previous section.

Software upgrading is an area that can cause friction between supplier and user. In some multiplexer and packet switched systems, upgrades may be as frequent as every six to twelve months and can contain some useful functions, especially in the network management area. A problem can arise when the customer sees little benefit in the upgrade compared with the work and risk involved in the migration. Suppliers, on average, will support the current and two previous releases. Even when they claim to support more, the quality of support may have fallen away. It is possible, therefore, to miss out a release. Unfortunately, there may be a hazard in doing this, where the manufacturers standard migration path to release n relies on n-1 being the current release. Normally, missing out a release is possible, but with the customer having to accept some extra risk. This friction need not arise if:

- the release/migration documentation discussed above is available;

- the supplier's migration policy is clear at the time the system is purchased. As with a number of other items, it will fall on the customer to find this out when he evaluates the system before purchase;

- with the benefit of the above information, migration plans can be made well in advance and properly budgeted for.

Moving users between networks will involve planning:

- to support two networks for the period of migration;
- to provide some redundant equipment to switch users between networks;
- establishing the unit of migration, for example moving one access line at a time rather than having to move all users in one go;
- for users to retreat to the old network if a problem arises that cannot be remedied in the time the migration window is open.

Some of this work would have to be done for a major network reconstruction.

Similar plans are required for moving computer centres, the difference being that the move should be paid for directly from the user's budget.

2.8 Costs and cost recovery

Costs

The following list of items covers the main capital expenditure (C), one-off installation charges (I) and running costs (R) which should be budgeted for:

EQUIPMENT	
SWITCHES	CIR
MULTIPLEXORS	CIR
MODEMS	CIR
PROTOCOL CONVERTORS, GATEWAYS	CIR
CABLING	CIR
RACKS	CI
NETWORK MANAGEMENT SYSTEMS	CIR
DIAGNOSTIC	
RECONFIGURATION	
LINE SWITCHING	
FAULT RECORDING	
PERFORMANCE ANALYSIS	
PROTOCOL MONITORS	
INVENTORY MANAGEMENT	
TRAFFIC ANALYSIS	

Note: Maintenance and Software Licensing are included in the above installation and running costs.

MEDIA	
DIGITAL TRUNKS (LEASED)	IR
DIGITAL AND ANALOGUE ACCESS LINES (LEASED)	IR
PSTN FOR DIAL-UP	IR
ISDN	IR
PUBLIC DATA NETWORKS	IR
LANS	CIR

PREMISES	
PREPARATION OF SITES (CENTRAL, NODAL, USER) FOR EQUIPMENT AND STAFF OCCUPATION	I
ENVIRONMENTAL EQUIPMENT, for example COOLING, POWER SMOOTHING, HUMIDITY	CIR
FIXED RUNNING COSTS, ie RENT, RATES HEAT, LIGHT	R

STAFF	
OPERATIONS	R
ADMINISTRATION	R
DEVELOPENT AND SUPPORT	IR
CONSULTANCY	I

Justification of the cost can be made:

- by comparison of the costs of providing a shared corporate network against providing discrete networks for each requirement. Availability and Response Time differences should be taken into account, and improvements in both of these should be expressed in terms of cost savings;

- by analysis of the year-by-year costs and savings over the development period. The results can be expressed using the usual financial analyses:

    - payback analysis,

    - cash-flow analysis,

    - return on investment factored using Net Present Value.

Cost recovery

All networking costs have to be recovered from income generated by the business. Network providers, within some organisations, aim to make their services self financing by some means of cost recovery. Ultimately, the aim is to recover costs from successful business enterprises to which the network has contributed. The major assumptions are that the network being developed will:

- provide for an overwhelming amount of the telecommunications requirements of the organisation;

- derive economies of scale from sharing resource;

- provide a service with high availability though a single, integrated management structure.

Initially, the provider will have to find, within the organisation, finance to start-up the development. Initial and subsequent developments should be financed from income, or borrowings against future income. The objective of cost recovery is to balance the cost of the service (including provision for developments) with income, a zero profit being acceptable. The income will be, almost exclusively, derived from the tariff applied to users for the service and the users, in turn, will have to include this revenue item in their budgets.

Tariffs

A fair tariff structure allocates costs against network resource used. Otherwise you could have the situation of the first user application having to pay for the network infrastructure and subsequent users only paying for local links from the network. As future requirements grow, the application that triggered the need for more infrastructure, would be penalised with that cost.

Ideally, the costs to the user will be based on a standard tariff which takes in to account the level of service required, number of locations, speed of access and resource usage. Volume of data is often used as a resource usage factor. Experience shows, however, that determining exact data volumes can prove difficult, especially compared with usage revenue generated. Far more costly is the physical provision and management of the links to give the required response time and throughput. Therefore, assessment on locations may be preferable. Location assessment can be categorised by link speed, host or terminal type of link, provision of dial-up, etc.

2.9 Implementation and management

Procedures

Implementation and management will tend to be accomplished by following a standard set of tasks, and will entail setting up and maintaining documentation systems. The implementation of communications networks can be very disruptive and organisations should be ready for this. Management of a network, however, should be almost transparent to the user. These tasks should be described in procedures which also establish responsibilities.

The major tasks are:

- Service Level Agreement (SLA) review;
- review of service performance;
- management of changes;
- incident/fault management;
- problem management;
- inventory Management.

The responsibilities for carrying out these tasks is largely an Administration function, with Operations providing the raw information for assessing service performance.

Service level agreement (SLA) review

The SLA is the formal contract of service between Service Provider and User and contains the following information:

- service provider;
- user;
- contact names;
- period of agreement;
- equipment interfacing to service;
- locations of user equipment;
- service cover time, for example 24 hours, 7 days per week; 9.00 to 5.00, 5 days per week. The cover may vary for different items of equipment connected to the service. For example, mainframe equipment would attract very high cover while terminals, which are only used during office hours, lower cover. Other factors may play a part in determining the cover, such as access to buildings out of office hours. The level of cover required should be examined very closely as it is a very costly revenue item;
- level of service. This will include availability (%), maximum occurrences of failures in a given period, time to restore service, standard down times for network changes and routine maintenance;
- escalation of problems. Time periods for escalation and titles of job holders involved in the escalation;
- procedures for changing the SLA and the schedules contained in it;
- other responsibilities, such as environmental conditions;
- charges and payments;
- software release information;
- limitation of liability;
- termination;
- disputes;

- force majeure;
- calculation of service time so that actual and contractual service performance, over a period of time, may be compared.

Review of service performance

Service should be measured in terms of the parameters listed under Level of Service and should computed from data collected over the review period. This will require a fault logging system and an inventory system as well as a performance analysis system. Clearly all of these parameters, as well as response time, have to be measurable. They are standard parameters required for measuring the performance to each user. Being able to monitor and compute performance should be considered as fundamental to the implementation of a network service.

Management of changes

This will be governed by one or more procedures covering the connection of new user equipment, movement of equipment and removal of equipment. It will detail the information to be provided and the standard schedule for implementation.

Incident/Fault management

Incident/Fault management procedures will cover what takes place when a user has a problem with the service. They will describe who is responsible for reporting a fault, where it is reported, who deals with it, details recorded and escalation. Detailed fault histories should be kept for future analysis, for example performance analysis.

Problem management

Problem management procedures are required to deal with:

- problems experienced by the user with the way the service is being provided. For example, network service maintenance personnel turning up unannounced at user premises; noise or heat problems with network equipment on user premises;

- problems experienced by the provider with the way the service is being used. For example, in one X.25 network, when a user's mainframe was not operating, call requests were being made every 30 seconds from 300 terminals and being cleared. Every one was being recorded on the network management statistics files which caused file management problems.

The procedures will describe how problems should be recorded and who is to take responsibility.

It is important to allocate responsibilities for user liaison, within the service organisation. Accountability is then clear to the user and within the service organisation. The objective in producing procedures and defining responsibilities is to structure a quality controlled service. The user should, ideally, have a structure with which to interface.

Inventory Management

The purpose of Inventory Management is to provide information on:

- all equipment and lines;

- equipment locations;
- equipment connections (for example site drawings, network drawings);
- rack layouts;
- floor plans;
- ancillary equipment (for example cooling, power smoothing);
- asset values;
- maintenance costs;
- line rentals;
- maintenance records;
- renewal dates;
- payment dates;
- times covered (for example 9-5 M to F);
- maintenance organisation & contact;
- escalation schedule & contacts (own and maintainers);
- site contacts and access details.

Inventory should also cover software and release information as listed in the section on Testing and Test Networks.

Quality Assurance

The aim should be to provide a quality system supported by procedures and with properly allocated responsibilities, based on recognised quality assurance standards (e.g. BS5750). There is a need to:

- produce fully documented procedures for development and management of the network services;

- negotiate and document Service Level Agreements with all customers, which are reviewed annually;
- review service performance with each customer at least quarterly, based on calculated availability versus agreed availability and other critical factors;
- have properly documented Services Contracts with suppliers and review annually;
- review performance of suppliers systems and equipment based on calculated versus contracted availability and on other critical factors;
- audit suppliers at least annually to establish that agreed codes of practice are being followed, for example equipment labelling, engineers identification;
- document and action non-conformances, e.g. customer complaints, accidents, damage, claims, late arrivals;
- internally audit adherence to procedures;
- hold management reviews to deal with all matters arising from the management of quality.

3
Products and Services

3.1 Background

The purpose of this chapter is to provide examples of packet switching products and services. There is no attempt to cover the products and services exhaustively as this would be an almost impossible task and the information would be out of date by the time of publication. New products are coming on to the market every day to provide PAD (Packet Assembler/Disassembler) functions, protocol conversion and gateways from LANs (Local Area Networks) as well as the normal switching and X.25 functions. The examples used, however, are representative in so far as functions and facilities are concerned. Recent standards have also influenced the emergence of new products, for example, X.25 dial up (X32), Frame Relay (I122), especially as the facilities offered can be developed in existing hardware. What is important, however, is that the choice of system is based on suitability for a user's business needs.

The process of assessing needs and evaluating requirements can present difficulties to many users and is an area where specialist help can be of great value. Any user is recommended to seek independent advice before committing himself to a particular supplier. The advisor chosen may come from the user's own staff or from outside the business, but should be independent of affiliation to any supplier. He must possess:

- analytical ability to turn the business requirement into a technical requirement;

- a broad knowledge of technology, its future direction and technical skills to construct alternative development strategies;

- business skills to assess the benefits of the alternative solutions against the costs and tactical skills to assess the risks involved in implementation. He/She must be able to provide the information in a way that is structured to allow a sound business decision to be made;

- organisational ability to specify the structure required to implement and manage the chosen solution. This should include defining the responsibilities for providing user service and the information that is required from the system for its management.

The management of the initial implementation is another area where independent specialist skills are required. Managing a company network on a day to day basis also requires skilled operators and administrators, who may not already exist within the organisation. One option to overcome this is Facilities Management. This is a service that is likely to be increasingly offered by service and product suppliers which tend to have the right sort of technical and administrative skills at their disposal. To obtain this type of business, they have shown willingness to take on the management of the customer's existing computer and communications equipment.

The step beyond this is to use third party networks such as the existing PDNs (Public Data Networks). These and any other networks with national coverage are ideally placed to take advantage of this market. Examples of other such networks are Istel, IBM's Managed Data Network as well as some of the large financial networks. The advantage to the supplier is that, having set up a network, the cost of management should level off, assuming that it is built to the right level of availability and the other important management criteria are taken into account.

One advantage to users is that it can help to overcome the problem of having to recruit skilled staff who are in short supply and whose skills are not those of the main purpose of the business, for example, financial services, retailing or manufacturing. It is often assumed that using a third party network is prohibitively expensive, but this may not be so in every case if all relevant costs are considered. In some countries, a third-party network may be the only option.

Users are recommended, however, to seek independent advice before committing themselves to a particular supplier to ensure that:

- the requirements are clearly specified and that the proposed solutions are evaluated in a structured way;

- there is a properly negotiated service level contract, which includes procedures and costs for changes;

- the performance of the service received is analysed based on criteria agreed between the parties, in advance, and that there are periodic reviews.

Users are advised to bring in the skills required to implement these controls and set up the structure for performance reviews.

Having put these controls in place, high levels of skill may not be permanently required.

It is important for users to take advantage of one other benefit, which is the independence that using products that conform to standards brings. They provide:

- interoperability between systems;
- higher vendor productivity as they only need to concentrate on the development of standard protocols and not devote resource to different protocol conversions;
- opening up of competition from other vendors who can now interconnect;
- simplified procurement as conversion costs and possible development delays should not occur, so comparison can be made on performance and cost.

3.2 Examples of services and products

Public Data Networks

Supplier

British Telecom
Network Services
Box 1351
London NW2 7HZ

Name of service

Global Network Services

Background

British Telecom have been providing Public Data Networking services for over ten years. They started in 1977 with the Experimental Packet Switched Service (EPSS). The Packet

Switched Service (PSS) followed this, along with the management of Euronet, which provided initial services in Europe. Euronet closed down five years ago as national services in Europe were implemented. More recently, BT have acquired Tymnet, the US PDN, and the rest of its worldwide packet network. The packet networks are marketed under the banner of Global Network Services (GNS). This in turn is part of Managed Network Services (MNS), which includes Cardway, GNS, Telecom Gold and GOLD400.

Cardway provides the means of connecting card verification terminals (for example, in petrol stations and shops) to authorisation mainframes.

GNS consists of PSS, Tymnet and IPSS. The core hardware employed in PSS is Telenet TP4 series switches (which are supplied, in the UK, by Plessey-Telenet). There are about 280 of these switches in 43 Packeting Switching Exchanges, interconnected by 48kbps lines. They are configured as a physical hierarchy, the higher layer providing trunk switching only. The lower layer provides access for X.25 users. There are also connections to other hardware (ie non-TP4) which provides SNA/SDLC, BSC and asynchronous support. It is a design constraint that five is the maximum number of switches that a packet is routed through. High availability of the core is provided by a combination of:

- component redundancy in the switches (including power supplies) and 1 for n port redundancy;
- switch and line redundancy using alteratively configured routes through the network;
- the internal proprietary network protocol which will switch routes between end user access points, transparently to the user, in the event of an internal network problem (see also OSI Transport Layer).

The interface to other network hardware is X.25, which means that the end-to-end resilience may be reduced because of the absence of a transport level protocol between the network end points.

IBM interfacing is provided, mainly, through the TP3 series concentrators. The services are SPAD (SNA/SDLC) and BPAD (BSC). SPAD will support host connections at 48000bps, all others are at 9600 and below. It is possible to provide IBM support on TP4s using a later release of software than that presently used in PSS. Dial back up is also available.

Asynchronous support is possible on TP4s but the line cards are prohibitively expensive. Various other equipment is used including Telematics switches which are used for the Vascomm network and for Dialplus. Dialplus differs from BT's earlier Dial X28 in that the connection from user equipment to the switch is error protected using MNP or V42 techniques, provided by the modem equipment. Note that asynchronous, leased line connections to PSS are not error protected.

High availability in the access part of the network can be maintained by call redirection, routeing to an alternative exchange, Dial back up and contracting with BT for a high level of maintenance cover.

The charging structure contains several elements:

- installation charge;
- connection rental;
- volume charge;
- duration charge;
- facilities charges.

Naturally, the different services attracted different charge element combinations and until recently, the X.25 leased line service attracted all five charge elements; the Duration charge has now been dropped. IBM service consists of one host and several terminal connections to form, in effect, a permanent virtual circuit. It therefore attracts the first two charges plus the fifth if any back up is required.

Tymnet is a large PDN in the USA. It also extends to about twenty other countries. It should not confused, however, with other national networks using Tymnet switches. These do not belong to BT. Additionally, the Tymnet Global Network is connected to a large number of PDNs worldwide.

IPPS is a large Bell switch which connects PSS to a large number of PDNs worldwide.

Range of services

Name	**Protocol**	**Interface**	**Speeds(k)**
Leased line			
X.25 Dateline	X.25	V24,V35,X21	2.4,9.6,48,64
SPAD	SNA/SDLC	V24,V35	2.4,9.6,48 (Host only)
BPAD	BSC	V24	2.4,9.6
X28 Dataline	X3/X28/X29	V24	1.2
Dial line			
X.25	X.25,X32.	V24	2.4,9.6
X28 Dial up	X3/28/29,	V24	1.2,2.4
Dialplus	MNP,V42		

Protocol Conversion, for example, X.25-Async, SNA-BSC, SNA-Async and BSC-Async.

Installation and support

Given the basic requirements, BT will provide a network design to satisfy the user. Quite naturally, this design will be based on BT services and products. Any user is recommended to seek independent advice before committing himself to a particular supplier, as discussed at the beginning of this chapter.

Network Management

Facilities for users

Access to the BT Help Desk is available up to 24 hours a day, 7 days a week, depending on the level of service required. New configuration and configuration changes are performed regularly to a strict timetable, to a set of properly documented procedures. The changes themselves are properly documented and checked.

An *insight terminal* is available to users, if required. The terminal, located at a user's site, provides real time information on the status of the parts of the network that affect the user.

Management by BT

Fault management

The procedure is for the Fault Reception Centre to take initial details of the fault from the user and to arrange for someone to ring them back within twenty minutes. During this time, the database is examined for customer details and any fault history.

Responsibility for the fault is allocated to:

- Tymnet faults to Luton Network Management Centre (NMC).

PSS faults to London NMC or Manchester NMC depending on where the fault is thought to be located.

Network monitoring

The Telenet network management system is capable of reporting losses of equipment and lines. The information is captured by the switches and sent to the central management system.

Call related information is gathered in the same way. This information, in BT's case, is used for accounting purposes and measuring utilisation of the network. The Colour Monitoring system provides real time network utilisation information using its layered colour maps.

Test calls are made into the network to continuously check the performance of the packet network and the Public Switched Telephone Network which connects dial-in users.

Software upgrades

BT undertake very thorough testing before migrating the network to a new software release. The reload has to be done across the network from the central Prime network management system, which is time consuming in a network of this size. If there is a problem which entails regressing to the previous release, a reload from the central system is required. BT tend to take releases some time after they are available, which should ensure that the bugs have been found on other networks.

Performance analysis

Statistics on hardware and media performance are collected and passed on to the relevant maintenance organisation for action. For PSS, they will be passed to other service organisations within BT.

BT's ServiceCare sets targets for installation and repair times with charge reductions if these are not met. For most users the measure of performance is actual service availability against designed service level. This is not provided by BT at the moment, although the information from which the figures can be derived is available. Users should also keep their own records of availability.

Security

This is performed at three levels:

- physical, by restricting personnel access to network sites;
- network access, by passwords and Closed User Groups;
- management Control, by audits and the independent Security Division.

Supplier

Mercury Communications
1 Riverbank Way
Great West Road
Brentford
Middlesex TW8 9RS

Name of Service

Mercury Data Network Services

Background

Mercury market two services for packet switching, Mercury Data Network Services (Mercury 5000) and International Packet Data Service (Mercury 5100).

Mercury 5000

Mercury 5000 is the national network which is based on a core of Northern Telecom switches interconnected using the Mercury fibre network. An access layer is also configured which provides customer connection services, for example:

- X.25;
- videotex;
- asynchronous connections, which can include MNP error protection;
- card authorisation;
- synchronous connections.

This access functionality is provided largely by Telematics switches. There is resiliency in the network through:

- redundant switches;
- redundant components in the core switches;
- transport layer functions, in Northern Telecom and Telematics equipment, which can provide end-to-end re-routeing in the event of network failure. Note that this functionality will only exist where both ends of the connection are on the same manufacturers equipment.

The charging structure contains several elements:

- installation charge;
- connection rental;
- volume charge;
- facilities charges.

Mercury 5100

Mercury 5100 is the international service. Mercury is part of the Cable & Wireless group, and operate an international packet network based on Tymnet equipment. The Mercury international gateway uses Tymnet equipment to connect to worldwide Public Data Networks. There is also a connection to British Telecom's Public Data Network. The Cable & Wireless 'Tymnet Equipment' network should not be confused with the Tymnet PDN which is operated by British Telecom's Global Network Services.

Range of services

Name	**Protocol**	**Interface**	**Speeds(k)**
Leased line			
X.25 Leased line	X.25	V24,V35,X21	2.4,4.8,9.6,19.2 48,64
Synchronous	SNA/SDLC	V24,V35	2.4,9.6
	BSC	V24	2.4,9.6
Asynchronous *see* Note 1	X3/X28/X29 MNP,V42,	V24	0.3-2.4 (inc 1200/75)
Dial line			
Asynchronous *see* Note 1	X3/28/29, MNP,V42, Tulsedata	V24	0.3-2.4 (inc 1200/75)

Note 1. The Viewdata Service can be accessed, asynchronously, at up to 2400bps leased line and at 1200/75 dial up. Error protection is provided by modems supporting MNP, V42 or Tulsedata techniques.

Telecommunications Facilities Management

This is a service offered by Mercury by which, from Mercury premises, the management of all or part of customers communications requirements can take place.

Installation and support

As with BT, Mercury will provide a design to satisfy a users requirements and the level of support for installing and running. The recommendations for seeking independent advice, that were advocated in the British Telecom section, should be also followed for Mercury.

When a connection is installed Mercury will test up to the user interface.

Network Management

Facilities for users

Maintenance support is available 24 hours, 7 days a week. Between 0800 and 1800 Monday to Friday, initial diagnosis is undertaken at the Network Control Centre. Outside of these hours, faults are handled by Tehnnical Support. They have identical equipment to the Control Centre. This centre also acts as the standby centre.

Configuration changes are performed to a schedule. The changes themselves are properly documented and checked.

Management by Mercury

Faults/Monitoring

The switches are capable of detecting line and equipment loss.

Alarms are sent back to the central site. There are facilities for V54 loopbacks, status checking, channel and packet tracing and V24 interface monitoring.

Configuration

Northern Telecom and Telematics equipment are relatively straightforward to configure, the routeing mechanisms being relatively simple. The changes are made on the central management systems and then sent across the network.

Software

Upgrades are subject to detailed testing before going live. Upgrades and reloads are sent over the network for loading on to the switches.

Information

Call related information is collected from the switches. This is used for billing purposes. The utilisation data is used for managing network capacity. Service Level Agreements are made with the customer and performance is reported to the customer.

Security/Access

Access to the network is controlled by user names, passwords and service names. Access to hosts/terminals is controlled by similar methods.

X.25 Switch Suppliers

Supplier

Telematics International Ltd
Isis House
Reading Road
Chineham
Basingstoke
Hampshire RG24 0TW

Background

Except in a few instances, Telematics do not sell directly to users, nor do they provide first line support, leaving this responsibility to the direct suppliers. Telematics supply services such as marketing, bid support, second line support, technical support and training.

The switching equipment is supported by a single network management system. Protocol and other functional support is fairly good because the product is marketed in a way that encourages suppliers to add value. The system provides a software environment platform which has enabled third parties to develop a range of functions including wide protocol support as well as protocol conversion. This particularly flexible feature makes the product attractive, but also creates a problem.

Communications products tend to have much more stable software than end user computer devices, and because they support a simpler set of functions, changes are less frequent and software development is under the control of a single group. This should mean that the support required for software is relatively low. This Telematics strategy means that there is a possible reduction in stability, and the likelihood of this are how long a product has been in use, how widespread is its use and how frequently it is changed. As with all requirements, users need to be able to assess the risks and benefits of a solution against the risks and benefits of other solutions.

Switches

The switches fall into two families: Access Communication Processors (ACP) and Programmable Communications Processors (PCP).

ACPs

These are lower end devices with three models in the range; ACP10, 20 and 30, with the following characteristics:

Ports	6-32
Speed	64k on up to 4 ports
Interfaces	V24, V35, V11
Protocols	X.25, SNA, Asynch
Throughput	50pps (128 bytes)
Local storage	Diskette or non-volatile memory

Resilience

Calls may be rerouted when a link fails and an alternative route exists.

X.25 dial out

This feature can be used as the primary link or an alternative for primary link failure.

Multipoint

This feature allows multipoint lines to be configured between ACPs. This obviously gives a saving in modems and line cards and, possibly, lines costs. There are also savings to be made in connection charges if the ACPs can replace other equipment on an existing multipoint network.

Address Conversion

This feature is normally required to connect the private X.25 network to a Public Data Network.

Upgrades

The units are field upgradeable by changing the Personality Module, which defines functions, e.g. protocols and memory size. New releases of software are upgraded in a similar fashion.

Network Management (SmartView)

Faults

Faults are detected by the switches and alarms sent back to the central management system. Also available is V54 loopback and a Line trace.

Commands

Link level Up, Down, Stat, etc., on X.25 and SNA links.

Monitoring

Trace for channels, packets, PU, LU etc., and status of EIA interfaces.

Configuration

May be built on the SmartView PC and stored, for each node. Down loading can be scheduled to occur at any time.

Information

Statistics and billing information may be collected from the nodes for the user's own analysis.

Security

NUI, CUG and various passwords. For SNA, pre-assigned LUs and pre-determined hosts can be restricted to use by to specific end users.

PCPs

PCPs are aimed at satisfying the needs of larger networks. This is the range of products, referred to in the Background section, that are rich in protocol support functionality. There are five models in the range; the S240, the Series 200, 600, 4000 & 5000, with the following characteristics:

Ports	32-448.
Speeds	64k on 12 to 32 ports. A 2 Megabyte frame relay interface is scheduled for the end of 1990. It is based on a line interface for the models at the higher end of the range.
Interfaces	V24, V35, V11

Protocols	X.25, SNA, Async, Ethernet, BSC plus a number of others developed by third parties, for example, X400, TCP/IP, parts of OSI.
Throughput	500-2000pps.

Users should exercise caution with throughput figures obtained from suppliers. They will generally be based on a single packet traversing a switch, however there should be other constraints placed on throughput figures. A factor of 40-60% utilisation should be applied depending on whether the end user connection is supported by binary synchronous, X.25 or SDLC protocols. It is also wise to halve the result on the assumption that acknowledgement will not be piggy-backed onto data packets, ie will occupy their own packet.

It is important for users to establish the basis on which the throughput figures are given. The responsibility for guaranteeing throughput should be put on the supplier by specifying the volumes and characteristics of the information that is to be handled by the network.

Resilience

At the component level, the only redundancy that can be provided is by configuring switches adjacently to allow sharing of line cards. The internal network protocol will switch routes between end user access points, transparently to the user, in the event of an internal network problem (see also OSI Transport Layer).

Address Conversion

This feature will allow the private network to be connected to a Public Data Network.

Network Management

The Telematics software product is called Interactive Networks Facilities (INF). It will provide management services for the ACP and PCP products. INF runs on a PCP, centrally or distributed on several PCPs throughout the network. A VDU is required for the user interface; printer and disk storage should also be configured. A Graphics Workstation (GWS), based on a Sun workstation is also available. All of the INF functionality, using graphical facilities is not yet available. However, a VT100 emulation window will allow full INF management to be performed from the GWS. A similar windowing method could be possible to manage other network entities, eg modems, from the GWS. INF has the following features:

Faults/Incidents

Faults are detected by the switches. Alarms may be sent to any of the Network Management peripherals, VDU, printer, disk. They are grouped by product and split into categories. The alarms can be filtered by category and severity. Alarms cover, for example, Network User Security, file security, node availability, link availability.

Commands/Monitoring

These provide details of a node and link status (packet and frame level), up and down links, redefining of connections.

Configuration

Configuration/reconfiguration of a node can take place at the Network Control Centre and downloaded to a node. A node will be off-line for up to 4 minutes while reloading.

Upgrades

Software upgrades are distributed using the same method as for configuration.

Information

Stats and Billing information may be collected from the nodes for processing by the user.

Security

NCC (Network Control Centre) access to Network Management system:

- Terminal. Established at configuration time to define access to the NCC as local, dial-in or X.25.
- User ID/Password access.
- Command Privilege/Menu access.
- Access to certain commands and menus.

Network access

- Telematics Access Products are available for controlling access from Async, SDLC (Synchronous Data Link Control) or BSC (Binary Synchronous Communication protocol) devices.
- CUGs (Closed User Groups) and passwords can be used.

Supplier

Olivetti
17-29 Sun Street, London EC2M 2PO

Background

Olivetti are suppliers of the BBN Communications range of packet switches. BBN were the developers of ARPANET, one of the first and one of the largest packet switched networks in the world. It is used by the US Military. Olivetti will design, implement and maintain networks for UK customers, BBN being responsible for software support. The system supports a good range of protocols and there are some very good management support tools.

Switches

Until recently the range comprised 3 switches, the C/3, C/30 and C/300. When supplied by Olivetti, the C is replaced by PSP. There is now a new high end addition to the range, the C/3000. Marketed within the same range are the C/10 PAD, the C/7 and C/70 Network Operations Centres (NOC). The NOCs provide the same functionality for networks of different sizes. The characteristics of the system are as follows:

	C/3	**C30**	**C/300**	**C/3000**
Ports	12	44	64	113
64k	2	32	32	113
Internodal	12	14	14	14
Interfaces	V24/V35/V11			
Throughput (pps)	300	400	900	3000
Simultaneous VC's	512	512	1024	14000
Call set up rate (per sec)	30	30	50 appx.	50 appx.

Protocol support

X.25 on the switches. Synchronous and asynchronous protocols are supported through the C/10 (or PSP/10). Its characteristics are:

Ports	4-32 sync 8-64 async 1-2 X.25 to switch node
Speed	Up to 64k to node, load balanced Up to 19.2k to hosts/terminals
Interfaces	V24/V35
Protocols	Triple-X, Async, BSC, SDLC/SNA, HDLC, Uniscope plus a number of other synchronous protocols.

LAN interfaces are supported by the T/200 IPR (Internet Packet Router) which support Ethernet and Token Ring interfaces. It provides routeing functions for certain protocols; the T/200 using addressing information in the protocol unit (ie message) to route to the destination. Bridging can be used to connect two network end points where a particular protocol is not supported for routeing. There are three products in the family. The characteristics of the family are:

Ports	1-26 LAN 1-52 WAN (Wide Area Network, typically using analogue, digital, satellite or radio links to connect end users countrywide or worldwide).
Interfaces	WAN - for example, V35, E1(2048 Mbps), T1(1544Mbps) LAN - for example, IEEE (Institute of Electrical and Electronic Engineers) 802.3, 803.5 (ie Ethernet and Token Ring)

Protocols	WAN	- X.25
	LAN	Routeing - eg, TCP/IP, DECnet, XNS
		Bridging - eg, LAT, IPX, AppleTalk.

The T/200s have a central management system which is separate from the switch management system.

Resilience

The new 3000 is the only member of the range to offer component redundancy. This is on the System Resource Board and power supplies.

Network resiliency can be provided by building redundant routes through the network There is a Transport level protocol that will reroute calls, transparently to the user, if another route is available. The Transport protocol does not extend as far as the C/10.

Address Conversion

This is not provided on this range of equipment. A Dynatech Gateway is usually supplied if this function is required.

Upgrades/Maintenance

It is necessary to switch off C/3s, 30s, and 300s for upgrades and maintenance.

Fast Packet

There are plans to implement this on the C/3000s.

Network Management

Network Management is based on the Network Operation Centre (NOC) using a C/7 or C/70, depending on size of network. Up to 4 NOCs can be configured to provide hot standby with network status information sent to all, if required. Terminals may be distributed around the network. Additionally, up to 2 workstations with a layer by layer graphical display of the network, can be configured.

Faults/Monitoring

Looping facility. Status of:

- Interface;
- PSN (Packet Switching Node);
- PAD;
- Ports on screen in real time;
- Netscope (a remote datascope facility using a DOS based IBM PC. At the moment its use is restricted to monitoring SNA traffic on the far side of the PAD).

Configuration

Configuration is done on the NOC which then can be downloaded to the relevant Node.

Software

Reloading of software can be done locally, from the next node or from the NOC.

Information

Call accounting for billing. Statistics:

- throughput (packets, bytes) on:
    - trunks,
    - switches,
    - host accesses,
    - events,
    - outage.

Design tools

DesigNet is a design tool which is run by BBN on data provided from the target network as a 'health check'. It is, in effect, a capacity management tool.

Security/Access

Network Access System providing:

- user authentication ID, password, destination;
- administrator defined access control;
- centralised master database.

The NOC functionality is migrating to a Unix based system on DEC under the Enterprise Management Architecture (EMA).

Supplier

Plessey-Telenet
Norton House
Stewart Road
Kingsland Business Park
Basingstoke
Hampshire RG24 0PL

Background

Telenet is one of the largest Public Data Networks in the USA and worldwide. There are also a large number of Public Data Networks using Telenet equipment, as well as a large number of private networks. The largest Public Data Network in the UK, BT's PSS, is largely Telenet, which Midland Bank uses on its private network. The main item of equipment in the range is the TP4 series switch, which has developed from the TP4200 through the TP4800 to the latest TP4900, and which is marketed as a trunk and as an access switch. Multi-protocol access is provided by the TP8000 series switch. The switch is limited to 16 ports. It is also possible to provide switching in a TP8000 sub-network.

The TP8000 have their own inter-nodal protocol which provides transport layer services. So do the TP4s. The protocol between TP4 and TP8 is X.25; therefore there is a certain lessening of resiliency if, say, the source of a link is a TP4 and the destination a TP8. this problem is being addressed in a future software release by providing a common transport protocol for TP4s and TP8s.

There are two other older TP series devices, TP2500s and TP3s. The TP2500 was developed by Plessey and can still be found in use to provide X.25 access. The different members of the TP3 family will provide access from BSC, SNA and Asynchronous users. The protocol from both TP3 and TP2500 to TP4 is X.25 with no Transport Layer functionality. The following analysis covers the TP4900 and TP8000 series.

The network is managed from a TP5000, which is a Prime machine. During 1991, it is planned to migrate the on line management functions from the Prime system to a Unix-based Sun workstation system. A particularly good feature of the network management is the Network Colour Monitor ('Colour') which provides graphical, real time information on network status from its highest level down to the status of port cards.

Switches/Access Nodes

	TP4900	**TP8000**
Ports	10 to 2000 async 10 to 1024 sync	8-16
64k	1024	All
Internodal	1024	All
Interfaces	V24/V35/V11	
Throughput(pps)		
Trunk Access	10000	
Access	1200	400
Call set up rate (per sec)	125	
Protocol support	X.25,X75,X32. IBM BSC & SDLC are planned for the future. IBM protocols are already supported on earlier TP4s. Async, Triple-X.	X.25,X32,Async, Triple-X,IBM SDLC & BSC.

Resilience

The TP4 range offers component redundancy for CPU, memory, power, Arbitrator plus 1 for n port redundancy. There is no component redundancy on TP8000s. Network resiliency can be provided by building redundant routes through the network. The TP4s and TP8s have their own independent transport layer protocols providing for end to end resilience, so there is only a problem if a TP4 and a TP8 are at the opposite ends of a route.

Address conversion

This function has been provided on the TP4s since software release 4.0.

Fast packet switching

This can be provided at the moment by the TP7/II. The series ranges from a model with a maximum of two E1 (ie the Euopean standard 30 x 64k structured channel configuration) trunks, two E1 ports plus 60 voice and 64 data input channels to a *hub* model with 96 E1 trunks, 96 E1 ports plus 2880 voice and 1248 data channels. Other features are:

Resilience

There is 1 for n component redundancy on all hardware including buses and power supplies. Rerouteing can also be configured for the event of network failure.

Fault Management

Functionality exists to detect faults at component and link level and to invoke available redundancy, automatically. Telenet also plan to provide Fast Packet support on the TP4900.

Network Management

Faults/Monitoring

The main tools used are Telenet Processor Reporting Facility (TPRF) and the Network Colour Monitor. TPRF is used to gather all events which can be recorded on screen, printer and disc. 'Colour' is a graphical display tool which provides layer by layer monitoring of the network. The colours of the links and equipment change to reflect losses and changes in utilisation. Numbers of calls, numbers of packets and errors can be monitored in real time.

Configuration

This is a relatively complex job requiring skilled people to effect routeing table changes. Configurations are prepared on the Network Management Prime, and can then be downloaded to the TPs. To retain compatibility, 'Colour' maps may also have to be changed.

Software

Loading and reloading of TP8s can be done locally, if the disc drive is configured. Otherwise, like the TP4, they are loaded from the Network Management system.

Information

Information is gathered for billing and for statistical analysis purposes. There exists a network administration tool called the Network Management Information System (NMIS) which can be used for billing and for analysis of network utilisation.

Security/Access

Provided by passwords, user IDs, Closed User Groups. There is also a Prime based system called TAMS for authenticating users right to enter the network and to be connected to particular links.

Supplier

Case Communications
Caxton Way
Watford Business Park
Watford
Hertfordshire WD1 8XH

Background

Case have been suppliers of telecommunications equipment and systems for some years. They are part of the Dowty group of companies. Case will offer a range of services from simple supply of equipment to installation management of a system and any accompanying media. They also offer maintenance and technical support, and training services.

The packet switching equipment and peripheral interfacing equipment offers a wide range of functions. The equipment is managed from a single network management system.

Switches

The equipment falls into a single family called the Series 8000 which can be managed from a centrally based Network Management System, the Series 5000. Essentially, the products fall into three categories: Protocol Handlers, Concentrators and Backbone Switches.

Protocol Handlers

The 8160 handles asynchronous connections and the 82xxs, synchronous protocols. Protocols and other functions can be changed by inserting a different module, called a POP-PAC, into a unit. Characteristics can include multi-protocol handling.

The main features of the units are:

Ports	1 network and up to 8 terminal ports.
Speed	50-9200bps (inc 1200/75) to terminals and up to 64kbps to the network.
Interface	V24, V11
Protocols	X.25 (DTE or DCE), SNA(SDLC), BSC,ICL CO3, Bull VIP 7700/7800, Async (X3,X28,X29).
Throughput	50pps (128 byte). This is said to take account of 'ACKs'.
Local storage	Non-volatile storage on the 8160 for configuration.

Concentrators

The 8225s can act as feeders to a backbone packet network or can be configured to form a small packet network. The main features are:

Ports	5-9
Speed	Up to 64k on 1 port in the 5 port configuration.
Interface	V24, V11
Protocols	X.25 (DTE or DCE)
Throughput	50pps (128 byte)
Set up rate	40 calls/sec

Backbone packet switches

The 8400 and 8500 main features are:

Ports	Minimum 4 to 32 on 8000 and 64 on 8500
Speed	2.4-64kbps
Interfaces	V24, V11, V35
Protocols	X.25 (DTE or DCE)
Throughput	200-3200 data packets/sec. Switches can be built up to the maximum configuration in modules of 4 ports. Each one is a separate port and processor module capable of handling up to 200 packets per sec.
Resilience	There is a dual bus and dual power supply. Functionality exists to reroute when a link fails and an alternative route exists.

The Hunt Group facility exists, which allows several DTEs to share a single group address. A Hunt Group will not function beyond a single node. There is, however, Call Redirection available.

Address Conversion	This feature is available, which is normally required for connecting to another private packet network or Public Data Network.
Local Storage	There are dual floppy disks from which the software can be loaded.

Network Management

The 5000 Network Management System comes in three sizes. The 5800 and 5310 PC based systems are for small and medium sized networks and there is a 5350 system, based on a Sun workstation, for larger networks. All have graphical interfaces. Features are:

Faults/Monitoring

- status alarms are sent back to the management system;
- loopbacks and test patterns are available for fault isolation;
- there is a call trace facility on the 8400 and 8500s.

Configuration

Configuration is done at the management system and downloaded to the nodes.

Software

Software is loaded across the network.

Hardware

The units are field upgradable. Board replacement, on the 8400 and 8500s, can be done with the power on.

Information

Statistical information is collected from the nodes into a spreadsheet on the management system, for manipulation by the user. Some inventory facilities are available for recording equipment details.

Integrated Network Management

Management of Series 8000 equipment and the Case DCX message switching equipment can be done from the same 5350 system. Case say that they also plan to manage their modems and LANS from the 5350. A windowing facility could be used through which other units may be managed, via a VT100 emulation.

Security access

- access to the network can be restricted by user identities and Passwords;
- network users can be split into community groups which have access to a common set of services.

Supplier

Infotron Systems
Westminster Tower
Albert Embankment
London SE1 7SP

Background

Infotron are, almost exclusively, suppliers of data communications equipment and systems. They have been in this business for many years, particularly in the area of multiplexers. They will also provide maintenance support, technical support and training services.

The packet switching equipment has a wide range of functions and interfaces and can be managed from a single network management system.

Switches

There is a single family of switches called Streamline 25. The Streamline 25 is slightly unusual in that it can provide a circuit switching service in the same unit as the packet switch. Infotron market this separation by featuring the circuit switching as the way of handling SNA and synchronous protocols. From the user interfacing point of view, this would look no different from any other switch handling multiple protocols. Transmission of the non-X.25 data uses time division multiplexing (TDM) which will minimise and fix the delay across the network. The TDM method, however, places greater demands on bandwidth which is determined by the number and speed of user connections, as opposed to X.25 where bandwidth is determined mainly by volume of data to be carried.

In the family, there is also a multiprotocol PAD and concentrator, called the Streamline 2510, which can concentrate a number of data channels on to a single StreamLine 25 channel or any X.25 network, private or public. This, of course, provides a method of connecting the SNA and synchronous data through the packet switched route. The main features are:

Streamline 2510

Ports	Up to 32. Each one is individually configurable.
Speed	Up to 64kbps
Interfaces	V24, V11
Protocols	X.25 (max 64k), X75 (max 64k), SNA (64k), BSC3270/2780/3780 (19.2k), Burroughs Poll/Select (19.2),NCR (19.2), Bull VIP (19.2), Async X3/X28/X29 (19.2).

Streamline 25

There are 3 switches in the 25 range: the 30, 50 and 70.

	30	**50**	**70**
Ports (fully configured)			
Packet	200	400	800
Circuit	1000	2000	4000
Speed (max)		2Mbps	
Throughput			
Packet(pps)	500	1000	2000
Circuit(Mbps)	16	32	64

Interfaces	V24,V35,V11, 2Meg (for circuit-switched).
Protocols	X.25 and X75 in the packet switched slice. Any synchronous on the circuit switched slice as they are transparent.
Resilience	Most of the components can be made redundant. The ability exists to reroute transparently, in the packet switched network, in the event of a link failing and another route being available.
Address Conversion	This feature, normally required for connecting to another packet network, is available.
Local Storage	Disks for software and configuration loading.
LAN Gateway	This LAN/X.25 Gateway at 64kbps is a feature of an Infotron product called Commix 32, which is an Ethernet LAN communications server. Besides X.25, it also supports DEC LAT, TCP/IP and OSI protocols.
Frame Relay	Infotron say that they are planning to implement this through a product from their supplier, Netrix.

Network Management

This is based on an Apollo workstation which may be operated from a single or multiple sites. There is a graphical interface for configuration and monitoring. The network diagrams are automatically changed to reflect configuration changes.

Faults/monitoring

Different threshold levels can be set so that an item can be monitored for different events.

Configuration

Changes are performed on the management system and sent to the nodes over the network.

Software

Similar to configuration, software is sent across the network. The software is supported from the USA.

Information

Statistics and billing information is sent across the network from the nodes. For the Streamline 25 it can be held on disk and sent when the network is quiet.

Security access

There are multiple levels of password protection for access to the management system. Closed User Groups are available for the X.25/Triple X connections.

Supplier

ICL
Kings House
33 Kings Road
Reading RG1 3PX

Background

ICL are the suppliers of Northern Telecom packet switching equipment and systems. Northern Telecom have been established in packet switching for a number of years. They run a large Public Data Network (PDN) in North America, which also extends worldwide. A number of PDNs also use Northern Telecom equipment, including that run by Mercury in the United Kingdom.

A wide range of functions and interfaces are provided. The equipment is managed from a single network management system.

Switches

The equipment is from a single family called the DPN-100. The network is managed centrally and is split between operations and administration functions. Operations functions reside on the switches while administration requires a VAX or IBM processor.

DPN-100 models	**Access lines**	**Pkts per sec**	**Links to network**	**Trunks**
5	8-32	64	up to 8	
10s	8-32	64	up to 8	
20s	32-64	3200		up to 12
10	32-104	800	up to 16	
15	32-208	800	up to 44	
20	32-176	3200		up to 14
40	200-800	3200		up to 16
50	300-3000	3200		up to 44
500	2000-30000	30000		up to 20

Speeds (up to)	**User access links**	**To network from access nodes**	**Trunks**
	64k	192k 2048k(on 500)	192 2048(on 500)

Interface	V24, V35, V11, G703
Protocols	X.25, X75, X32, Asynchronous, X3/X28/X29, SNA/SDLC, BSC
Resilience	There is redundancy for most components. The routeing architecture will reroute data, when a failure occurs, without data loss.
Address Conversion	This has to be done by an external device.
Local Storage	There are disks for storage of software, configuration and collection of billing and statistical information.

Network Management

The real time element of management (for example, alarms, monitoring) is an inherent part of the switch functionality, requiring only an asynchronous terminal interface to the operator. Network administration tasks (for example, configuration) are undertaken on a VAX (or IBM) processor. In common with other suppliers, NT say that they are planning to migrate the administration system and 'real time interface' to a workstation system. In this case, it is a Sun system and, of course, will have a graphical interface.

Faults/Monitoring

Status alarms are sent back to the central site operator. A number of diagnostic tools are available, for example:

- V54 loopback;
- Interactive protocol tester;
- DPN packet trace.

Configuration

Configuration is performed on the Network Administration System (NAS). They are sent over the network and can be held on the local disks.

Software

Software is also sent over the network and can be held on the local disks. Support is generally for the past 2 years releases.

Information

Billing and statistical information is collected on the local discs for transmission, over the network, to central storage. The NAS processes the information to provide utilisation and delay statistics, for example.

Security access

Access to the management system is restricted by user identity and password. User access is restricted by passwords and Closed User Groups.

Supplier

Timeplex
Langley Business Centre
Station Road
Langley, Slough
Berkshire SL3 8DT

Background

Timplex is another organisation whose business is almost exclusively in telecommunications products and systems. The core of their business has been in multiplexers, where they were one of the earliest manufacturers. They are now a part of the Unisys organisation. Their move into packet switching is more recent, reflecting the widening of this market.

Timeplex will offer a range of services from installation to Facilities Management. They also offer maintenance support, technical support, and training services. In addition there is available disaster provision insurance under which Timeplex will bring a new node into operation within 48 hours. The equipment has a wide range of functions and is managed from a single network management system.

Switches

The equipment is in a single family called Timepac, consisting of 3 different size switches, providing similar user functions, 2 synchronous PADs and an asynchronous PAD.

Switches	NP100E	NP500	NP1000
Ports	20	28	28
Speeds			
Sync (up to kbps)	64	128	128
Async (bps)	50-38400	50-38400	50-38400
Throughput (pps)	100 (128)	500 (512)	1000 (512)
Call setup rate (per sec)	30	60-100	120-200

Interfaces	V24, V35, V10, V11
Protocols	X.25(DTE or DCE), Asynchronous, X3/X28/X29, SNA/SDLC, BSC
Resilience	Redundancy for the main processor module is available Rerouteing around a failed component will take place automatically and without data loss, where an alternative route exists.
Address Conversion	This is available for translating addresses, which is required when connecting to other packet switched networks.
Local Storage	Available on all 3 systems for software and configuration.

Synchronous PADs (TS10 and TS40)

Ports	1 network and 1(TS10) to 4(TS40) user.
Speeds	Up to 19.2kbps for the sync protocols and 64kbps for X.25.
Interfaces	V24, V11
Protocols	Burroughs Poll/Select(sync and async), Sperry Uniscope, Honeywell (or Bull) VIP 7700 and 7800.
Resilience	The PADs will attempt to re-establish a lost connection so long as the terminal remains active, with no loss of data.

Asynchronous PAD(TA48)

Ports	1 or 2 network(X.25) and 8 to 48 user.
Speeds	19.2kbps (1 port) or 9.6kbps (2ports) on the network side and 9.6kbps (Async) on the user side.
Interfaces	V24.
Protocols	X.25, Asynchronous, X3/X28/X29.
Frame Relay	Timeplex say that they are planning to incorporate Frame Relay on to their multiplexing equipment.

Network Management

An integral network management system resides within the switches. Management information and functions can be interfaced to through an asynchronous terminal. Several management sessions to each node can be supported simultaneously. The switches also support a report generator for sorting, displaying and printing network statistics. Also available is the Time/View 2500 Element Management System. This is based on a Sun (80386 CPU UNIX) workstation, with a graphical interface. It does not support the TS PADs.

Faults/Monitoring

Status alarms and information can be brought back to the network operator interface.

Configuration

Configuration is relatively easy. After initialising changes, the switches exchange information and automatically generate network routeing tables. Changes can be performed at the central site and sent over the network.

Software

Software can be sent across the network for loading on to the nodes or via the local storage. There new releases about every 3 months. To avoid having to upgrade all nodes at the same time, Timeplex claim that the network will run with different versions of software in different nodes.

Hardware upgrades

These can be done in the field with the power on.

Information

Billing and statistical information is available for analysis.

Security access

- the switches can be physically locked;
- access to the Management system is restricted;
- network access can be restricted by user name and password;
- called destinations can be also protected by password;
- closed User Groups can be also used.

Integrated Network Management

- The Time/View 2000 management system, for the Link multiplexers, can gateway through NetView/PC to IBM's NetView. Alarms can be sent through and monitoring performed.
- Time/View 2000 provides an interface to AT&T's ACCUMASTER Interrogator. Alarms are passed through and the operator can enquire on the network for further information.

Supplier

Racal-Milgo
Landata House
Station Road
Hook, Basingstoke
Hampshire RG27 9PE

Background

Racal-Milgo have been a presence in the telecommunications market for a long time, particularly data communications. They offer a full range of equipment and systems.

Their support includes maintenance, technical and training services. They will provide a network implementation service for their own and third party equipment and media. They will also provide a facilities managed service on a full or part time basis. During the last 3-4 years they have been implementing the Government Data Network (GDN), which is designed to interconnect equipment run by a large number of Government departments. The network is owned and run by Racal Data Networks. Racal provide a similar service for Government voice communications.

Racal have been supplying packet switching equipment in the UK for about 7-8 years. They supplied BBN equipment, most notably to NatWest Bank. About 5 years ago they changed to supplying the Telematics range of switches. The functions provided by these switches and their network management are described in the Telematics profile. It is Telematics policy to supply to users through third parties. Racal, with their extensive technical and maintenance resources, should be ideally placed as end user suppliers.

Racal say that they are active in the developing of network management integration. They are involved in the OSI network management committees as well as developing interfaces into IBM's NetView, AT&T's Accumaster And Enterprise Management Architecture (EMA).

Case studies

I - ICL home workers

Purpose

The case being described is for connecting home workers to ICL mainframes in Manchester, Bracknell and Reading. The main purpose of the network is to connect the community working exclusively at home to development systems. This part of ICL exists to develop mainframe systems (VME), using VME mainframe facilities. The working population at the mainframe centres is large. Just over one hundred work exclusively at home.

Background

The community that works exclusively at home, developing mainframe systems, has existed for some time. Initially they used various ICL DRS terminals. They were the binary synchronous type of terminal connected by dial up lines through the Public Switched Telephone Network (PSTN), at 2400bps, to Manchester. ICL CO3 was the line protocol used. Following on from this there was some research carried out on how terminal equipment, supporting asynchronous communications, could be connected. One of the first developments was for the BBC Model B. This in resulted the production of a chip that provided emulation of ICL Communication Mode in the terminal equipment.

The problem of handling the screen presentation of non-ICL terminals was overcome by VME (i.e. mainframe software) development. This provides protocol conversion of the character strings that control screen presentation on the terminals in the network. Originally, this handler was called the Sponsor which has now been developed into a standard product called the Flexible Terminal Handler (FTH). This will support all the standard screen handling functions. Any new terminal can be supported by adding, into FTH, the character strings information required for the particular terminal's screen manipulation.

Most recently, ICL have provided the community working exclusively at home with MS-DOS PCs. They have packaged it with a 2400bps asynchronous modem and a printer. Also in the package is a software product from Softklone called MIRROR III, which provides asynchronous terminal emulation. Within MIRROR III, ICL has developed code, called ASTI, to communicate with FTH and to invoke the call set up routines. The terminals are also able to access the Electronic Mail services used throughout ICL. (*see* Figure 15).

New network development

In 1989 ICL set themselves the objective of reducing call charges in this network. BT's PSS Dialplus service was about to be launched at this time. They had already looked at the existing X28 Dial-up service. The problems for ICL with this service were:

- it is limited to maximum speed of 1200bps;
- the PAD distribution meant that a proportion of users still have to use trunk calls to connect. Coverage, according to BT, is about 72% of PSTN users;
- the dial-up connection is not error protected.

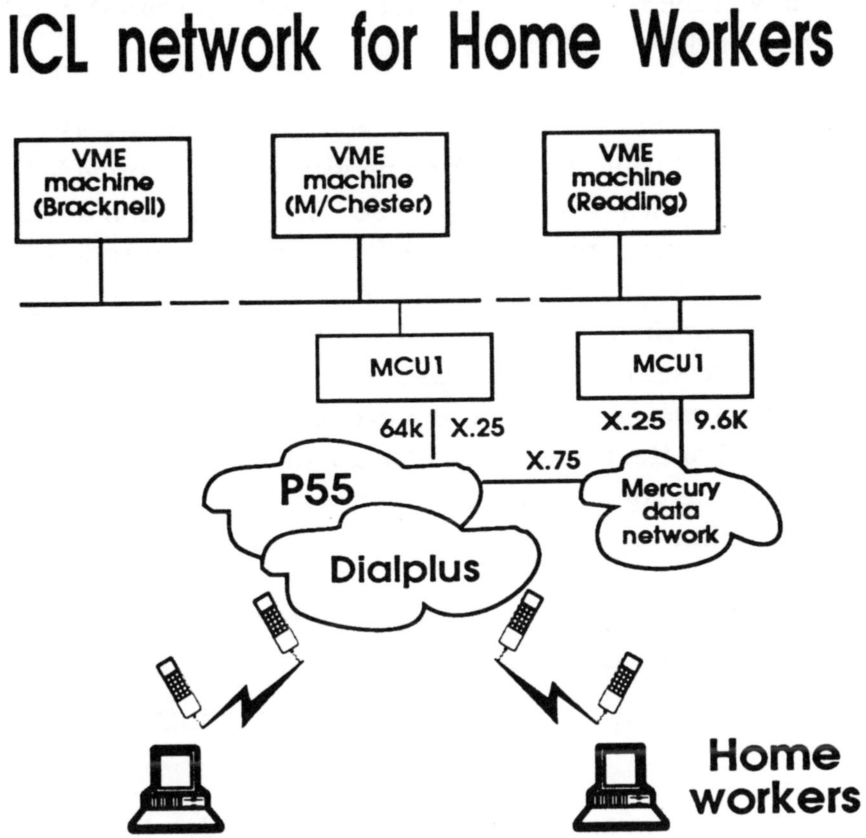

Figure 15 ICL network for home workers

These problems were overcome by Dialplus which offered:

- up to 2400bps;
- a PAD distribution that aimed at 93% of possible users being connected by local calls. ICL's population has ended up with 98%;
- error protection of the dial-up link using MNP. The V42 service supports MNP classes 2-4 and LAPM;

- no volume related charges for calls set up from the dial-up end of the connection.

The BT field trial for PSS Dialplus started in January 1990 and ICL participated in it. The 100 home workers were provided with connection facilities by February 1990. The connection at the ICL development centre end is a PSS X.25 Dataline connection, at 64 kbps, into the Manchester centre. The cost control on this line, for the home workers, is simple. There are no call related charges as all calls are set up from the user end. There is the connection charge, of course. After this there are only the recurring charges for the rental of the Dataline and for about 100 Logical Channels. Users requiring connections to other centres (i.e. Bracknell and Reading) are linked through existing ICL inter-centre connections.

The X.25 line is physically attached to a MCU1 FEP device which is linked by the OSLAN local area network to a VME mainframe. All these items are ICL products; OSLAN is the ICL proprietary LAN.

There is a degree of resilience for the 64k link. If this is lost, calls are redirected across the PSS to Mercury Public Data Network link. There is an existing Mercury 'Dataline' connected to the ICL network. The speed of this connection is 9600bps and there are only 30 Logical Channels available. Monitoring has shown that 30 is about the number of Logical Channels being used during the day on the BT dedicated link and it is running at 64k. Some degradation of service, therefore, might be expected.

Recently, ICL launched a special offer to staff to enable them to buy the ICL manufactured PCs. This has led to about 200 other workers connecting from home, on a part time basis. Mostly, this is during evenings and week-ends.

Management and Security

Management of this network is relatively simple. It has been simplified because there is no longer a requirement for banks of dial-up modems, PSTN hunt groups and FEP connections at the development centres. These have been replaced by the single 64kbps Dataline.

The access security is provided by the password feature within the Dialplus service. It is possible, of course, for users to make connections to other PSS end points at ICL's expense. ICL ask users to sign an undertaking that they will only use the service to connect to ICL. The usage details, however, are available on line and can be examined for unauthorised connections, by the ICL system supervisor.

There are fault finding diagnostics developed in the PCs. This consists of logging, on the PC screen, the stage a call has reached in its set up. The information is a valuable diagnostic aid when a fault occurs.

Accounting information can be accumulated by recording the times of setting up calls and clearing calls.

Benefits

The use of PCs as the terminal equipment has enabled connect time to be kept to a minimum by use of certain system development aids and working practices:

- documentation and some software development can be done off line using the PC facilities;

- disc to disc file transfer is done using simple asynchronous communications that formats the file for display. The file can be then printed or screened off line;

- the call set up routines can be invoked automatically from the PC.

Overall, the service has provided ICL with the means of achieving their objective of reducing line costs which are down by about 20%.

Problems

The problems and concerns they have with the service are:

- there is no user operated facility for changing the NUA associated with a PSS end user service name on the user's Dialplus menu. The name of the service is listed on the user's Dialplus menu and call set up is started by keying the number of the service on the menu. The change facility is required if the NUA of the end user service has to be changed. At the moment, the Dialplus user cannot change the NUA associated with a particular service name on the menu. BT would have to perform the change and would charge for doing so;

- they are counting on a development of BT's Guardian product to provide increased access security;

- the limitations on how passwords are amended, in the system, makes changing passwords difficult;

- another problem with passwords is that bills are associated with a password. If passwords are changed at any time other than the end of a billing period billing information for a user will be in several places;

- BT's call set up interface must not change without warning because the ICL scripts that automate call set up would not function;
- the option of 9600bps, for some users, would be welcome.

II - Strathclyde Police

Purpose

The development of Information Technology in such areas as crime administration and office automation has been a strategic aim of the Strathclyde Police Force. A suitable communications network is seen as an important foundation on which to build IT.

Background

The Strathclyde Police have been users of Information Technology since the mid-1970s. They implemented one of the first police Command and Control systems in the country. A Command and Control system is used for reporting incidents, e.g. as reported by the public, and for allocating resources to these incidents. Users in Force Control, Special Departments, Headquarters of Divisions and Sub-division offices access this system from 121 VDUs and 65 printers. There are about 55 offices.

They have also used systems for Personnel and Licensing as well as using the Local Authority system for Statistics. They have had developed a HOLMES system. HOLMES evolved in the aftermath of the 'Ripper' enquiry. These systems are used for handling the data collected during major enquiries. There is a standard specification issued by the Home Office which describes the functional requirements, allowing any hardware platform to be used. In Strathclyde's case, the system was developed on McDonnell-Douglas equipment. This system is accessed by asynchronous terminals.

In the mid-1980s, an internal Force Working Party identified a number of areas of development which would be of benefit to the Force:

- crime administration system;
- office automation;
- replacement command and control system.

In addition, a structured communications network was seen as a key foundation for the Information Technology strategy. The network in existence at the time had been built up in an ad hoc manner as a star of lines from Force Headquarters radiating to Force offices. With an already high investment in communications it seemed sensible to see if the network could be structured in such a way as to keep future development simple and achievable at a low cost.

Network Development

After consultation with a number of bodies such as the Directorate of Telecommunications at the Home Office it was decided to introduce an integrated voice and data network (*see* Figure 16). Until this time little regard had been paid to the part that a voice network could play in the Information Technology infrastructure. The traffic was measured and analysed and, from this, potential savings were identified. These savings were to be realised in

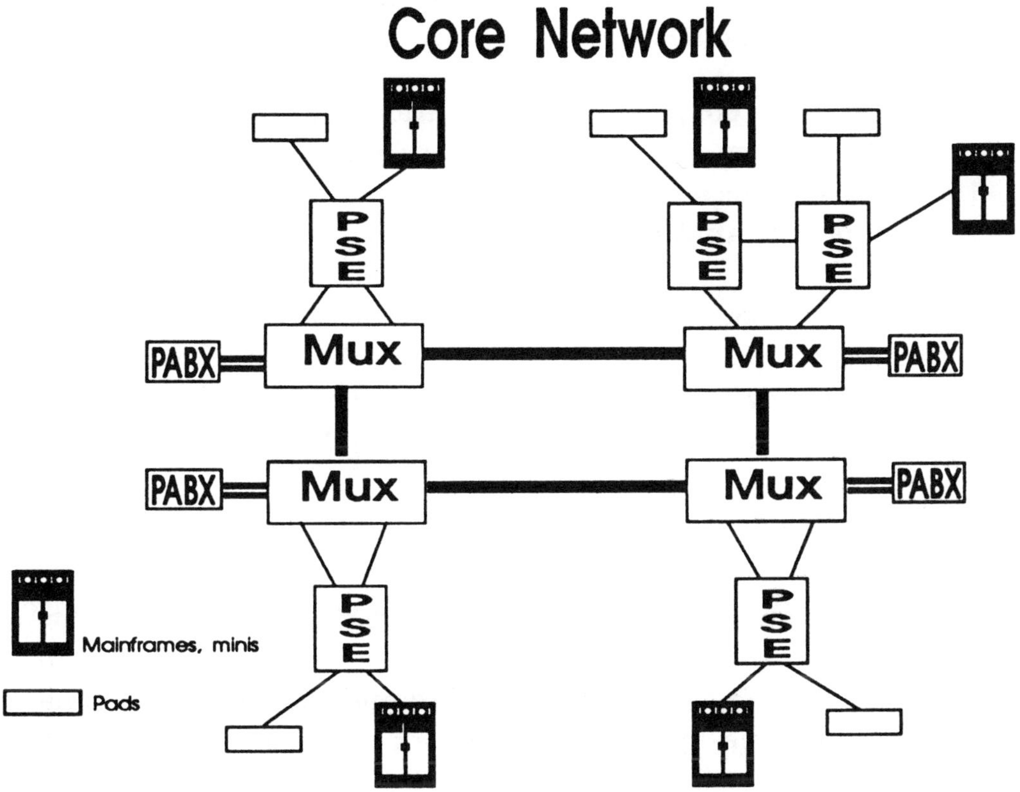

Figure 16 Core network

building a voice network with a high speed (2Mbits/s) digital communications backbone. This would provide flexibility for growth. Further cost savings could be made by sharing the digital backbone with the data traffic.

The voice network would be run on digital exchanges, which it was decided to locate in Force Headquarters at Glasgow and

Divisional Headquarters at Paisley, Motherwell and Kilmarnock. The switch chosen was the Plessey Integrated Switched Digital Exchange (ISDX), which interface using the Digital Private Network Signalling System (DPNSS).

The switches were purchased from British Telecom and the lines rented from them. The voice network was introduced in April 1989 and 40 offices are now connected to the switches by British Telecom private analogue circuits.

The user response to the improved service has been very enthusiastic. Public Switched Telephone Network (PSTN) usage charges have dropped dramatically. The Force is keeping usage under close scrutiny using Call Logging equipment.

The network for data was required to fill the following objectives:

- it was not to add unduly to the computer systems' response times;
- it would provide flexibility for interconnecting a wide range of computer developments;
- it would follow international standards (i.e. OSI).

After evaluation of the tenders the Case 8000 series X.25 network was chosen. Case 7410 multiplexers were also chosen to provide the means of sharing the 2Mbits/sec circuits between the voice and data switches. This core data network was introduced in April 1990.

Management and support

Support of the network hardware, software and media, when faults occur, is by the suppliers. Operational Control and other management tasks are undertaken by the Force.

The Force already had extensive experience of managing networks, through their large radio network and with the terrestrial analogue links used by the existing computer systems. There is also the Call Logging equipment, mentioned above, for monitoring voice network usage. The Case network management system is also available and has a number of diagnostic and administration features.

Benefits

- voice;

The large savings in PSTN charges have prevously been mentioned.

- data;

- **Home Office (HOLMES)**

The HOLMES system, for handling the data collected during major incidents, is intended to be used from anywhere. In practice this means Police premises with the right accommodation. An Incident Room will typically be equipped with 6-8 terminals for HOLMES access. The system capacity would allow 6-8 Incidents to be handled concurrently, although it would tend to be less because of training requirements. Previously, setting up 'incident' terminals, in certain locations, could be frustrated by lack of availability of links on the private telephone network. Now they can be set up wherever there is a data network end-point.

- Office Automation;

Office Automation is in quite extensive use throughout the Force. A major use has been for The Crown Office Pilot Project in which the data network has played a large part. When the police in

Scotland believe that an offence has been committed, they pass a report and other information to the Crown Office where a Procurator Fiscal undertakes any prosecution required. The Police prepare reports on their Bull DBS6 minis which can be accessed through dumb terminals and PCs attached to the network via a LAN. The information is keyed in by data preparation staff. An important breakthrough was for the various Procurators Fiscal involved to agree to standardisation of reports. This has enabled the preparation and modification of the document to be handled through a Standard Generalised Markup Language. There is pressure to prepare reports as they have to be out for court appearances. The network is used to pass the reports, in X400 envelopes, between the police Bull machines and the Procurators Fiscal. Information from the Scottish Criminal Records Office (SCRO) on previous convictions can also be merged with the reports.

These computer based records have provided the Crown Office with a very effective case tracking system. There is further interest in using the network to pass the reports and other case material from the Crown Office to the Sheriff's Court. This would complete the final side of the information square shown in Figure 17. The side not previously mentioned is the Court inputting details of convictions to the SCRO system.

- Command and Control system;

A replacement Command and Control system is due to be handed over to the Force in the next few months and will be accessed through the data network. One of the biggest problems to be overcome is user training. 675 key users have been identified for training. Normally, the training would take about a week and cater for a batch of 10 people. By using the network to provide access from additional locations, 55 people can be catered for at a time and the training completed in a minimum of 13 weeks.

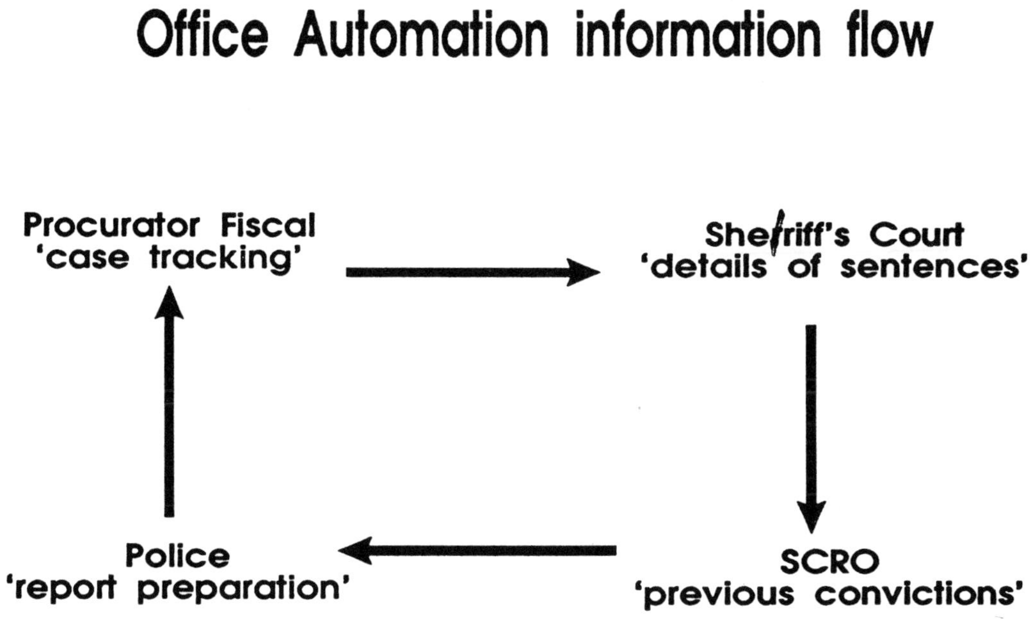

Figure 17 Office automation information flow

- Other systems;

Future plans, it is believed, will have a large dependency on the use of PCs connected through LANs to the network. Other systems planned which would use the network are Crime, Administration, Personnel, Licensing and Vehicle Fleet Management.

The use of the network (*see* Figure 18) is still fairly new, especially as far as the data part is concerned. Some of the expected gains have materialised, for example, the reduction of PSTN call charges. Some have come about partly through having a switched network in place, for example, The Crown Office Pilot (Office Automation) Project.

III - Retail financial organisation

Purpose

The case being described is for connecting counter terminal equipment, printers and automatic teller machines (ATM) to mainframe systems. There are backup mainframes, at a separate site, which attach to the network to take over the processing should the need arise. The system exists to store and process customer account information. The information originating, largely from the counter terminals in the 650 branches nationwide and from the ATMs. Files are moved daily to the backup site and there is a weekly check to see that the network can connect to it.

Background

The counter terminals and ATMs are on logically separate networks. There has also been a large degree of physical separation. The network is undergoing a large amount of change while the central systems and end user equipment remain intact. The two systems are linked together by a serial communications line between the central site equipment. The two networks were brought together physically by the use of 40+ multiplexers located at various nodal locations. The multipoints were connected in to the multiplexers at the nodes and fanned out to the respective

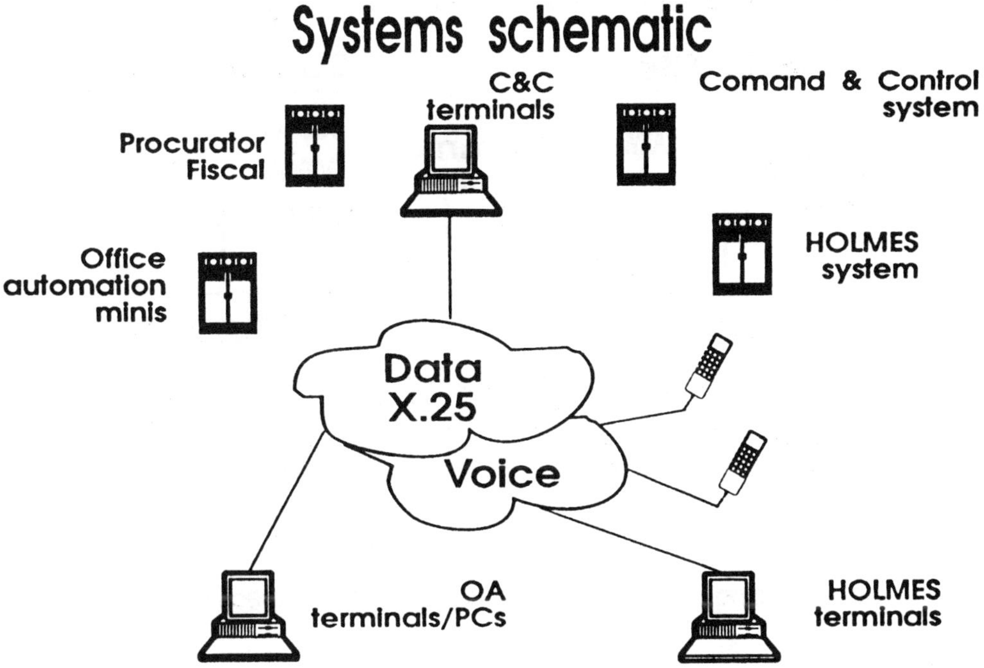

Figure 18 Systems schematic

the multiplexers at the nodes and fanned out to the respective central site processors (*see* Figure 19).

New network development

The organisation has now installed a packet switched network from Olivetti based on BBN equipment (see Products section).

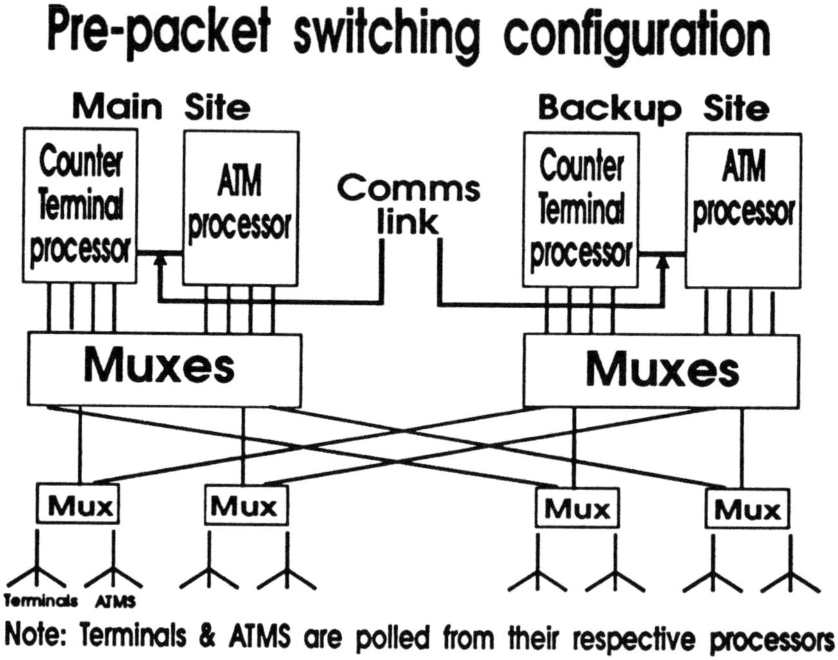

Figure 19 Pre-packet switching configuration

There is a configuration of 96 switches; 8 x C/300s, 15 x C/30s and the rest C/3s. The Counter Terminals have largely moved to the new network and the ATMs are due to follow. When the migration is completed, there will be a single 9600 bps from each branch to the network. This will support Counter Terminals and any ATMs in the branch (*see* Figure 20).

There has been a recent upgrade to the Olivetti LSX branch controller to provide X.25 support; the Counter Terminal equipment is connected into the branch controller. Similarly, the ATMs will be upgraded to support X.25. To achieve the objective

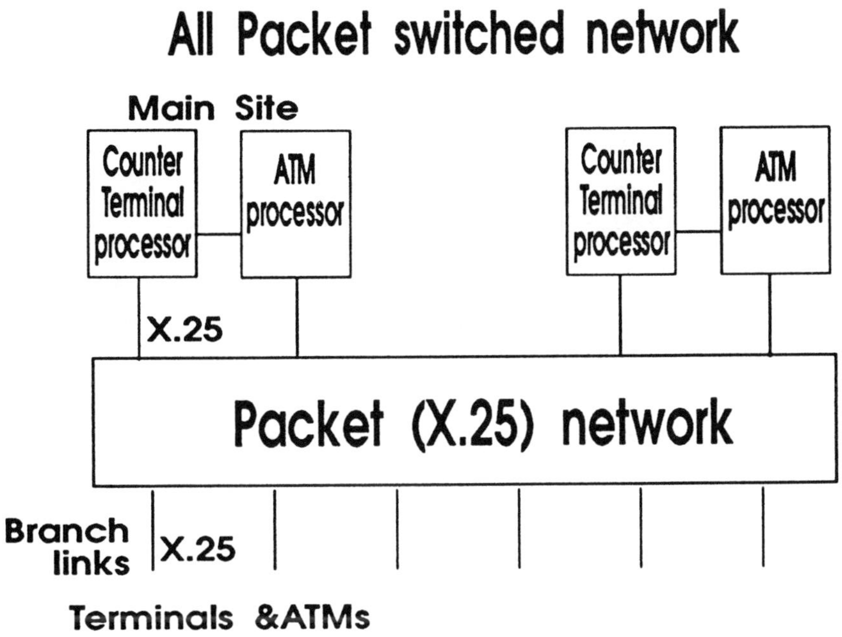

Figure 20 All packet switched network

of a single line from each branch, a branch concentrator device is being used. On the branch side, there will be a port for the concentrator connection and a port for each ATM. Another X.25 port will then connect the branch to the network. The lines from the branches are two-thirds digital.

They are BT Kilostreams. BT were unable to supply digital for the remaining connections so they have supplied a service to resemble Kilostream with similar parameters. There is no requirement for dialled backup.

At the central site, the connections are being simplified by the introduction of an X.25 based network. Only a few 64kbps links are required from the central site switch to the Unisys equipment. It will be similar for the Stratus when the ATMs are connected (*see* Figure 21). There is a similar configuration at the backup site. Moving the network on to the backup site is done by dynamic changes to the network configuration.

Management and support

Supplier support for the network is during branch opening hours. Support is from the suppliers of the equipment and media. Operational management uses the multiplexer managementsystem, the modem management system and the packet switching network management system. There is also Netscope, from BBN, which is a protocol analyser for ports on the network. It can also monitor round trip delays. Performance and utilisation analysis on the Switches is available. Other network administrative systems are being planned.

Benefits

- cost savings on revenue are expected from moving to a single network;
- there is flexibility for growth and connections;
- the flexibility opens up the possibilities for connection of other applications.

IV – Greenalls

Purpose

This case study is of a network for connecting terminals located in 550 Greenalls Public Houses to their mainframe computer site

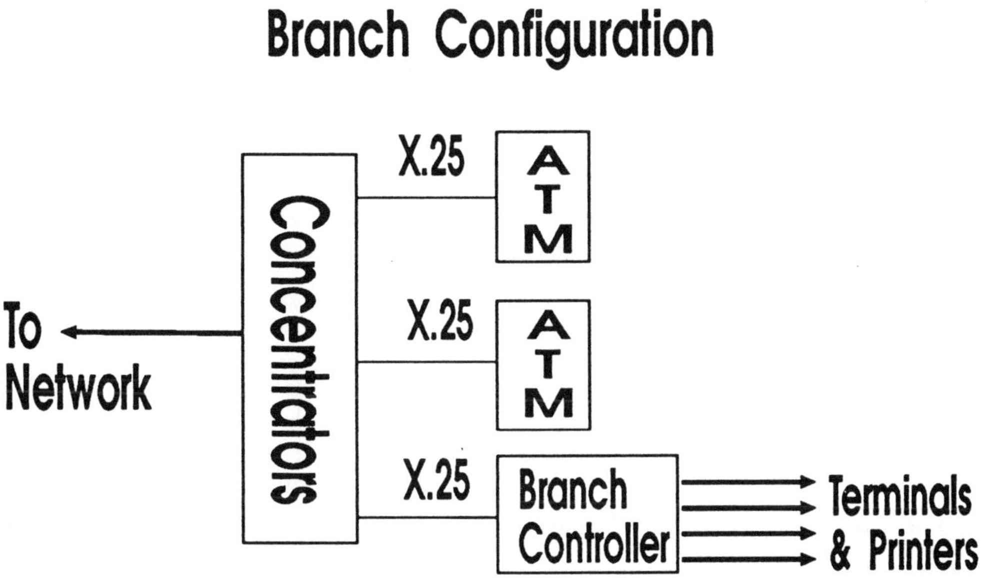

Figure 21 Branch configuration

in Warrington. The most important use of this network is for control of costs. Every day, details of income are entered from each Pub and, weekly, details of expenditure. Receipts are sent in by post. There is also a facility available for sending messages over the network. The sales management hierarchy can also access the system to monitor the costs.

Background

The system is based on Videotext technology. Simple Videotext terminals are installed in the pubs. At the computer centre, the data is collected on a IBM6150 RISC processor. The information is later transferred to an IBM AS400. The pubs are located in the north-west of England extending up to Cumbria in the north, North Wales in the west and the Midlands in the south. Location is particularly important in this network. The brewery has done some detailed research into alternative means of connecting as it can have a significant effect on costs. They are minimising the costs by using a mixture of British Telecom dial-up and the Mercury Public (packet) Data Network.

To handle communications they use a FEP from BT This provides the interface for the asynchronous dial-up connections at the asymmetrical speed of 1200/75bps. It handles the X.25 interface for the Mercury PDN connections. It also deals with the Videotext protocol, stripping it off before passing the data across to the IBM processor. There is also a hunt group of 29 telephone lines and these are attached to auto-answer modems plugged into the FEP.

Networks

Dial-up directly to the computer centre is principally for users whose calls would not attract the b-line (ie 56kms) tariff; 'b' is the charging category as shown in the Telephone book. The users' terminals are programmed with the telephone number of the hunt group, which is dialled from a key depression.

The other pubs, lying in the b-line area in relationship to Warrington, are programmed to dial a Mercury Videogate number (*see* network diagram in Figure 22). There are 80-90 Videogate machines nationwide so that connections can be made that only attract local call charges. This, of course, is the attraction for Greenalls.

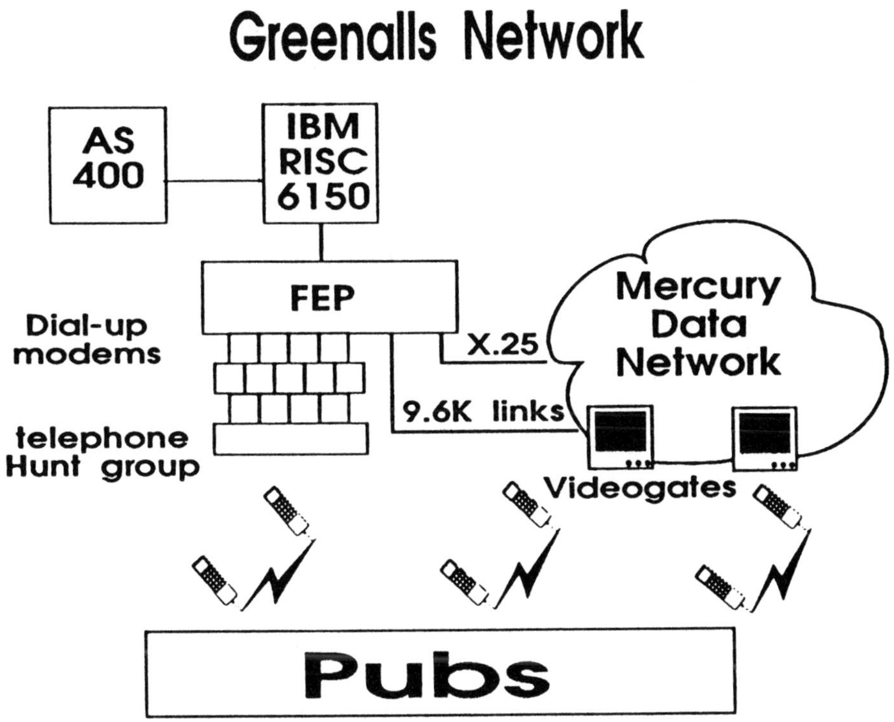

Figure 22 Greenalls network

The Videogate (formerly known as a Microscope) machine is supplied by GPT and supports up to 30 ports. It uses standard Videotext protocols to the terminals and across the network to the computer centre FEP.

There is one problem with this method of connection in that response times are slower than direct dial up. The problem is not of the network's making but lies in the way that the application handles communications. The characters are echoed from the central site equipment. The echoing serves to enable users to check for errors on the unprotected dial up line. Echoing can be done from the Videogate, but this would also involve amendments to the central site system because of its involvement in handling screen changes. The terminals are also programmed with the hunt group number for backup.

Management and support

This is essentially a simple network. The centre, when faults are reported to them by users, look to the carriers, BT and Mercury,

to diagnose the problem. Monitoring is done by examining the bills for call charges and network usage.

Benefits

- the availability of Management Information is far faster;

- when Payroll is linked up it will reduce the turnaround by about one and a half days;

- there is a reduction in the number of staff required for data preparation and checking;

- call charges are reduced; there are fixed overheads for rental of the two X.25 links into the computer centre and the charge for making use of the Mercury Data Network. There were initial connection charges and there are duration charges on the Mercury network. These can then be compared with the call charges saved. Savings will also be made when another 115 Pubs, in the Midlands, are taken on-line.

Glossary

ASCII — *American Standard Code for Information Interchange*, also known as IA5 (International Alphabet number 5) by CCITT. An 8 bit code, including 1 parity bit, for representing printing and control characters.

Asynchronous — A synchronisation method where each character (usually 8 bit ASCII) is synchronised by a fixed duration start bit in front and a stop bit at the end, so that the receiver can determine the boundary between characters. Even parity is used by convention.

Bridge — A *Bridge* connects two remote LANs at the data link level (OSI level 2).

BSC — *Binary Synchronous Communication protocol.* An early IBM character-oriented synchronous link control procedure. Used in conjunction with IBM3270 presentation protocol, it was implemented on many systems besides IBM and used widely for interfacing.

CCITT Another name for the International Telegraph and Telephone Consultative Committee (ITTCC).

COPs *Character oriented protocols.* These are data link control procedures which use the ASCII control characters. They are polling protocols: examples are IBM's BSC, ICL CO3, VIP 7700/7800 and Burroughs Poll/Select.

CLNS *Connectionless Network Service* where data is sent on to the network without any call establishment phase.

CONS *Connection Oriented Network Service* where a call is established before data is sent. During call establishment, negotiation can take place on packet size, throughput and error rates.

CUG *Closed User Group.* An optional X.25 user facility which permits users belonging to the group to communicate with each other but not outside the group. This can be relieved a little by having CUGS with outgoing or incoming access.

DCE *Data Circuit-terminating Equipment* is CCITT terminology for network devices such as modems, packet switches, etc.

DTE *Data Terminal Equipment* is CCITT terminology for terminal devices, PCs, mainframes, etc.

EIA The *Electronic Industries Association* (US), which published standards for data communications.

FTAM *File Transfer Access and Management.* An OSI Application Layer standard for accessing and administering files on end user systems.

Gateways A *Gateway* provides protocol conversion at the equivalent of OSI level 4 and higher, between two different networks.

HDLC *High level Data Link Control protocol.* It is bit oriented and data transparent, ie it has no special control characters except the opening and closing 'flag'. This is unlike character oriented protocols. There is a polling class: Normal Response Mode; an example is SDLC. Asynchronous Balanced Mode is used in HDLC, where it is known as LAPB (Line Access Protocol).

I122 Recommendation for Frame Relay.

IEEE The Institute of Electrical Engineers (US), which has played a leading role in establishing Local Area Network standards.

ITTCC *International Telegraph and Telephone Consultative Committee.* A committee of the International Telecommunications Union (ITU), which is an agency of the United Nations.

LAN Normally a network within a building, not involving common carrier services. The speeds are relatively high (1 to 100+ Mbps) and error rates low.

LAN MAC — *LAN Media Access control.* The three most popular are:

- ISO 8802/3 or CSMA/CD (Carrier Sense Multiple Access/Collision Detect) which was developed from Ethernet;
- ISO 8802/5 or Token Ring;
- FDDI (Fibre Distributed Data Interface).

NSAP Address — *Network Service Access Point Address* identifies a particular location in a sub-network, i.e. a self contained network within a global network.

OSI — Refers to the ISO *Open Systems Interconnection model.* The model defines seven protocol levels. Standards are available for all levels and so are products based on them. Interfacing is between peers, i.e. level 7 in one end user system interfaces to level 7 in another.

PAD — *Packet assembler/disassembler.* It allows access to an X.25 network from asynchronous terminal equipment. Its operation is governed by the Triple-X standards.

Parity — An error detection technique where one bit is added to each character to make the number of bits always even (or odd).

PDN — *Public Data Network* - packet switched.

PSTN — *Public Switched Telephone Network.*

Router A *Router* connects remote LANs at the network level (OSI level 3) and can select network paths.

SDLC *Synchronous Data Link Control.* IBM's bit-oriented link control procedure. Identical in all major respects to HDLC; the only differences are in a few, little used, facilities.

SNA *Systems Network Architecture.* IBM's de facto standard for networking.

Synchronous A synchronisation method where characters are sent, contiguously, in blocks preceded by two or more synchronisation characters (ie ASCII SYN), so that the receiver can determine the boundary between characters. Odd parity is used by convention.

TCP/IP *Transmission Control Protocol/Internet Protocol,* which are part of the Internet protocol suite. The suite was promulgated by the US Department of Defence. TCP provides a connection oriented, error free transmission service similar to ISO Transport layer class 4. IP is similar to the ISO Connectionless Network Service.

Triple-X CCITT recommendations on X3, X28 and X29.

V24 List of definitions for interchange circuits between a DTE and DCE (modem). Generally applies to speeds of 19.2kbps and below.

V35 Data transmission at 48kbps using 60-108kHz group band circuits.

V42 Error correcting protocol for modems. It includes MNP classes.

V54 Loop test devices for modems.

VT *Virtual Terminal.* An OSI Application Layer standard describing common facilities for terminals.

WAN *Wide Area Network.* As opposed to a LAN, a WAN interconnects geographically remote sites. The link speeds are generally lower than LANs and they have higher error rates.

X21 General purpose interface between DTE and DCE for synchronous operation. Used for transmission at 64kbps.

X.25 Interface between data terminal equipment and data circuit-terminating equipment for terminals operating in the packet mode on public data networks.

X28 DTE/DCE interface for an asynchronous data terminal equipment accessing the PAD facility on a PDN.

X29 Changes for the exchange of control information and user data between a packet PAD and a packet mode DTE or another PAD.

X3 Packet assembly/disassembly PAD facility in a Public Data Network.

X32 Interface between DTE and DCE for packet mode terminals accessing a packet switched

PDN through the PSTN, ISDN or a public circuit-switched network.

X400 An OSI Application Layer standard for transmitting electronic mail between systems.

X75 Terminal and transit call control procedures and data transfer system on international circuits between packet switched networks.

Index